Analytical Methods for
Toxics Determination

Analytical Methods for Toxics Determination

Editors

Clinio Locatelli
Marcello Locatelli
Dora Melucci

MDPI • Basel • Beijing • Wuhan • Barcelona • Belgrade • Manchester • Tokyo • Cluj • Tianjin

Editors
Clinio Locatelli
Alma Mater Studiorum University of Bologna
Italy

Marcello Locatelli
University "G. d'Annunzio" of Chieti and Pescara
Italy

Dora Melucci
Alma Mater Studiorum University of Bologna
Italy

Editorial Office
MDPI
St. Alban-Anlage 66
4052 Basel, Switzerland

This is a reprint of articles from the Special Issue published online in the open access journal *Molecules* (ISSN 1420-3049) (available at: https://www.mdpi.com/journal/molecules/special_issues/Analytical_toxics_Determination).

For citation purposes, cite each article independently as indicated on the article page online and as indicated below:

LastName, A.A.; LastName, B.B.; LastName, C.C. Article Title. *Journal Name* **Year**, *Article Number*, Page Range.

ISBN 978-3-03936-716-0 (Hbk)
ISBN 978-3-03936-717-7 (PDF)

© 2020 by the authors. Articles in this book are Open Access and distributed under the Creative Commons Attribution (CC BY) license, which allows users to download, copy and build upon published articles, as long as the author and publisher are properly credited, which ensures maximum dissemination and a wider impact of our publications.

The book as a whole is distributed by MDPI under the terms and conditions of the Creative Commons license CC BY-NC-ND.

Contents

About the Editors vii

Preface to "Analytical Methods for Toxics Determination" xi

Nora Hamad Al-Shaalan, Imran Ali, Zeid A. ALOthman, Lamya Hamad Al-Wahaibi and Hadeel Alabdulmonem
Application of Composite NanoMaterial to Determine Phenols in Wastewater by Solid Phase Micro Membrane Tip Extraction and Capillary Electrophoresis
Reprinted from: *Molecules* **2019**, *24*, 3443, doi:10.3390/molecules24193443 1

Mirko Salinitro, Annalisa Tassoni, Sonia Casolari, Francesco de Laurentiis, Alessandro Zappi and Dora Melucci
Heavy Metals Bioindication Potential of the Common Weeds *Senecio vulgaris* L., *Polygonum aviculare* L. and *Poa annua* L.
Reprinted from: *Molecules* **2019**, *24*, 2813, doi:10.3390/molecules24152813 13

Florin Dumitru Bora, Claudiu Ioan Bunea, Romeo Chira and Andrea Bunea
Assessment of the Quality of Polluted Areas in Northwest Romania Based on the Content of Elements in Different Organs of Grapevine (*Vitis vinifera* L.)
Reprinted from: *Molecules* **2020**, *25*, 750, doi:10.3390/molecules25030750 31

Xinwei Wang, Shan Lu, Tianzheng Wang, Xinran Qin, Xilin Wang and Zhidong Jia
Analysis of Pollution in High Voltage Insulators via Laser-Induced Breakdown Spectroscopy
Reprinted from: *Molecules* **2020**, *25*, 822, doi:10.3390/molecules25040822 59

Li Han, Yue-Tao Li, Jin-Qing Jiang, Ren-Feng Li, Guo-Ying Fan, Jun-Mei Lv, Ye Zhou, Wen-Ju Zhang and Zi-Liang Wang
Development of a Direct Competitive ELISA Kit for Detecting Deoxynivalenol Contamination in Wheat
Reprinted from: *Molecules* **2020**, *25*, 50, doi:10.3390/molecules25010050 71

Li Lin, Wen Yang, Xing Wei, Yi Wang, Li Zhang, Yunsong Zhang, Zhiming Zhang, Ying Zhao and Maojun Zhao
Enhancement of Solasodine Extracted from Fruits of *Solanum nigrum* L. by Microwave-Assisted Aqueous Two-Phase Extraction and Analysis by High-Performance Liquid Chromatography
Reprinted from: *Molecules* **2019**, *24*, 2294, doi:10.3390/molecules24122294 85

Patricia Garcia Ferreira, Carolina Guimarães de Souza Lima, Letícia Lorena Noronha, Marcela Cristina de Moraes, Fernando de Carvalho da Silva, Alessandra Lifsitch Viçosa, Débora Omena Futuro and Vitor Francisco Ferreira
Development of a Method for the Quantification of Clotrimazole and Itraconazole and Study of Their Stability in a New Microemulsion for the Treatment of Sporotrichosis
Reprinted from: *Molecules* **2019**, *24*, 2333, doi:10.3390/molecules24122333 97

Yue Zhang, Yan Zhou, Shujun Chen, Yashi You, Ping Qiu and Yongnian Ni
Analysis of the Overlapped Electrochemical Signals of Hydrochlorothiazide and Pyridoxine on the Ethylenediamine-Modified Glassy Carbon Electrode by Use of Chemometrics Methods
Reprinted from: *Molecules* **2019**, *24*, 2536, doi:10.3390/molecules24142536 113

Angela Tartaglia, Abuzar Kabir, Songul Ulusoy, Halil Ibrahim Ulusoy, Giuseppe Maria Merone, Fabio Savini, Cristian D'Ovidio, Ugo de Grazia, Serena Gabrielli, Fabio Maroni, Pantaleone Bruni, Fausto Croce, Dora Melucci, Kenneth G. Furton and Marcello Locatelli
Novel MIPs-Parabens based SPE Stationary Phases Characterization and Application
Reprinted from: *Molecules* **2019**, *24*, 3334, doi:10.3390/molecules24183334 . . . . **123**

Ahmed Galal Eldin, Abd El-Galil E. Amr, Ayman H. Kamel and Saad S. M. Hassan
Screen-printed Microsensors Using Polyoctyl-thiophene (POT) Conducting Polymer As Solid Transducer for Ultratrace Determination of Azides
Reprinted from: *Molecules* **2019**, *24*, 1392, doi:10.3390/molecules24071392 . . . . **139**

About the Editors

Clinio Locatelli (Associate Professor, Analytical Chemistry) 1975: Degree in Chemistry from the University of Ferrara; discipline: Analytical Chemistry. 1977: Qualification to the Profession of Chemist. 1977–1980: Holder of a scholarship from the National Research Council (Rome, Italy). 1980–1992: University Researcher, Scientific-Disciplinary sector Analytical Chemistry, Department of Chemistry of the University of Ferrara. 1992–1994: Associate Professor, Scientific-Disciplinary sector Analytical Chemistry, Department of Chemistry of the University of Salerno. 1994–present: Associate Professor, Scientific-Disciplinary sector Analytical Chemistry, Department of Chemistry "G. Ciamician" of the University of Bologna. 1980–1996: Several study periods for a total of about two years at the Research Laboratory for Inorganic Chemistry of the Hungarian Academy of Sciences in Budapest (Hungary) and at the Department of Analytical Chemistry of the University of Chemical Engineering of Veszprem (Hungary). In these centers, both theoretical and application lines of research have been consolidated for some time, aimed at the possibility of carrying out voltammetric multicomponent analysis of metals, also and above all, in the presence of strong interference from the voltammetric signal. RESEARCH SUBJECTS—The research activity (author of more than 160 publications in peer-reviewed international journals, and more than 200 congress communications) is addressed to theoretical and applied problems. The main research lines are the following. – Study of adsorption processes of organic molecules and their determination at trace level. – Set up of voltammetric and spectroscopic analytical methods in the determination, at trace and ultra-trace concentration levels, of heavy metals in real complex organic and inorganic matrices, especially of environmental, food, industrial, cosmetic, and forensic types. – Theoretical survey addressing the optimization of the analytical parameters such as sensitivity, selectivity, accuracy, precision, and limits of detection in metal determination by voltammetric and spectroscopic techniques. TEACHING ACTIVITY—The teaching activity (1980 to date) has always been focused on courses related to the scientific disciplinary sector Analytical Chemistry (Analytical Chemistry, Analytical Chemistry Laboratory, Instrumental Analytical Chemistry, Environmental Analytical Chemistry). From the Academic Year 1995/96, until the end of its activation, he was teacher at the School of Specialization in "Chemical Control and Analysis Methodologies" in the course Analytical Methodologies I. Academic Years 1996/97–2001/02: Professor of "Inorganic Chemical Analysis Methods" in the Specialization Course in "Marine Pollution Control" at the Faculty of Mathematical, Physical, and Natural Sciences of the University of Bologna throughout its activation period.

Marcello Locatelli earned his degree in Chemistry at the University of Bologna, Department of Chemistry "G. Ciamician" with his thesis "Development of an Analytical Methodology for the Analysis and Identification of Protein Adducts by Mass Spectrometry". He earned his PhD from the University of Bologna, Department of Chemistry "G. Ciamician" with his thesis "Combined Analytical Methods of Mass Spectrometry for the Study of Impurities in Drugs and Metabolites of Biomolecules". He is currently Associate Professor of Analytical Chemistry at the University "G. d'Annunzio" of Chieti-Pescara, Department of Pharmacy. His research activity is devoted to the development and validation of chromatographic methods for the qualitative and quantitative determination of biologically active molecules in complex matrices from human and animal (whole blood, serum, plasma, bile, tissues, feces, and urine), cosmetics, foods, and the environment. This applies to the study of all processes related to preanalytical stages such as sampling, extraction and purification, separation, enrichment, and even the application of conventional and combined analytical methods for the accurate, sensitive, and selective determination of biologically active molecules. Recently, particular attention is focused on innovative (micro)extraction procedures like MEPS, FPSE, MIP, DLLME, and SULLE. These procedures have been applied to different compounds, from synthetic and natural origin (glucosamine, 5-aminosalicylic acid, natural or synthetic bile acids, anti-inflammatory agents, drug association and fluoroquinolones, secondary metabolites from natural sources, and heavy metals). In the development of methods, predictive and chemometric models are also tested both for the optimization of extraction protocols and for final data processing. Particular attention is given to new instrument configurations for the quantitative analysis in complex matrices. Locatelli is author of more than 157 manuscripts, 116 congress communications, 1 patent subject to approval, 13 book chapters, and 3 books, in addition to being Guest Editor for 13 Special Issues. His attested scientific activity based on Scopus (8th of June 2020) includes having a h-index of 33, 148 papers, and 2778 citations. In addition, he is a reviewer of the following international journals: *Analytica Chimica Acta, Current Bioactive Compounds, Journal of Chromatography A, Talanta, Trends in Analytical Chemistry* (a sample from more than 100 international peer-reviewed journals). He is a referee for MIUR Institution for National Projects (SIR) and included in the register REPRISE (Register of Expert Peer Reviewers for Italian Scientific Evaluation) in the "Basic Research" section. Locatelli served as referee for VQR (2011–2014) and is a referee for other universities of proposals through competitive tenders for the allocation of University funds for the activation of research grants. He serves on the Editorial Board of Molecules for the section "Analytical Chemistry" in addition to *Current Analytical Chemistry, Separations,* "Current Bioactive Compounds, *American Journal of Modern Chromatography, Journal of Selcuk University Science Faculty, Reviews in Separation Sciences, and Cumhuriyet Science Journal*. He is Associate Editor of *Frontiers in Pharmacology* section "Ethnopharmacology", Review Editor of Frontiers in Oncology section "Pharmacology of Anti-Cancer Drugs", and Review Editor of *Frontiers in Medical Technology* section "Nano-Based Drug Delivery". He is a member of the Scientific Committee of the journal *Scienze e Ricerche*, published by the Italian Book Association.

Dora Melucci obtained her Master of Science in Chemistry at the University of Bologna, with her thesis entitled "Determination of Polypeptides by HPLC". She then obtained her Master in Chemical Methodologies for Control and Analysis at the University of Bologna with a thesis entitled "The Gravitational Field-Flow Fractionation Technique (GrFFF). Fractionation and Absolute Quantitative Analysis of Particulate in Dispersion". Finally, she obtained her PhD in Chemical Sciences at the University of Ferrara with a thesis entitled "Characterization of Polymers by Means of Thermal Field-Flow Fractionation (ThFFF) Using Decalin as a Solvent". She has served as Researcher and Assistant Professor (branch Analytical Chemistry) at the Department of Chemistry "Giacomo Ciamician", School of Sciences, University of Bologna, since being appointed in 1999. Her research subjects have included the following. Separation Science: FFF of macromolecules in solution and dispersed microparticles; standardless and absolute analysis in ETA-AAS and HPLC; ThFFF of industrial polymers in collaboration with industry; flow-FFF and GrFFF of real samples (starch, yeast, and metal nanoparticles for biostatic materials); miniaturization of separation tools; cell sorting; combination of FFF with chemiluminescence; and development of multianalyte competitive immunoenzymatic methods using dispersed nano- and microparticles. In the framework of these subjects, she was Coordinator of the Local Unity of Bologna in a national project (PRIN) entitled

"Microfluidic Separation of Nanosystems". She started an autonomous research line in 2005 with the basic keyword chemometrics, in which she continues to be involved. The topic is application of chemometrics to the development and validation of innovative analytical methodologies, with a focus on direct and non-altering methods of analysis. The fields of application are food, environment, pharmaceutics, forensics, biotechnology, cultural heritage. The analytical techniques employed are AAS, NIR, voltammetry, Raman, GC, LC–MS, FT-IR, XRD. Up to May 2020, her work includes 76 articles, 20 book contributions, and over 100 communications in national and international meetings, corresponding to a h-index of 17. Her teaching activity in 2002–2020 includes teaching responsibility in Analytical Chemistry, Laboratory of Analytical Chemistry, Analytical Chemistry and Law, Principles of Quality and Safety (Bachelor in Chemistry); Chemometrics (Master of Science in Chemistry); Chemometrics for Forensic Analysis (Master in Forensic Chemical and Chemical—Toxicological Analysis). Part of her additional academic roles include being Local Coordinator for the University of Bologna of the National Ministerial Project for high-school student guidance (Scientific Degrees Plan), and Department Delegate for university student tutoring.

Preface to "Analytical Methods for Toxics Determination"

In recent decades, but especially in recent years, the rapid spread of toxic species has occurred in all matrices. Evidently, these contaminants, both of inorganic and organic origin, represent very dangerous pollutants for human health, owing to their bioaccumulation and toxicity. They result from almost all human activities, such as, for example, industries, agriculture, vehicular traffic, and urban heating.

Considering that often these species occur in the various real matrices at extremely low concentrations, the various analytical methodologies must evidently demonstrate their possible application by verifying their accuracy and integrity at all steps: sampling, sample preparation, instrumental measurement, and statistical data processing.

Another extremely important aspect concerns the fact that the development of new analytical methodologies and the contemporary lack or inadequacy of regulations regarding the determination of toxic species in the most varied matrices can obviously give, to the legislator, the possibility to update and improve the abovementioned regulations.

This volume aims to attract contributions on all aspects linked to the different analytical methods used for the determination of toxic species in the most varied matrices—food, environmental, forensic, biological, and so on—focusing particularly on the fundamental parameters of interest to set up an analytical procedure, such as precision and trueness (that together give accuracy), the limits of detection and quantification, selectivity, and especially sensitivity.

Clinio Locatelli , Marcello Locatelli, Dora Melucci
Editors

Article

Application of Composite NanoMaterial to Determine Phenols in Wastewater by Solid Phase Micro Membrane Tip Extraction and Capillary Electrophoresis

Nora Hamad Al-Shaalan [1,*], Imran Ali [2,3,*], Zeid A. ALOthman [4], Lamya Hamad Al-Wahaibi [1] and Hadeel Alabdulmonem [1]

1. Department of Chemistry, P. O. Box 84428, College of Science, Princess Nourah bint Abdulrahman University, Riyadh 11671, Saudi Arabia; Lhalwahaib@pnu.edu.sa (L.H.A.-W.); haalabdulmenem@pnu.edu.sa (H.A.)
2. Department of Chemistry, College of Sciences, Taibah University, Al-Medina Al-Munawara 41477, Saudi Arabia
3. Department of Chemistry, Jamia Millia Islamia, (Central University) New Delhi 11025, India
4. Department of Chemistry, P. O. Box 2455, College of Science, King Saud University, Riyadh 11451, Saudi Arabia; zaothman@ksu.edu.sa
* Correspondence: nhalshaalan@pnu.edu.sa (N.H.A.-S.); drimran.chiral@gmail.com (I.A.)

Academic Editors: Clinio Locatelli; Marcello Locatelli and Dora Melucci

Received: 29 August 2019; Accepted: 20 September 2019; Published: 23 September 2019

Abstract: Composite nanoparticles were used in solid phase micro membrane tip extraction and capillary electrophoresis to determine phenol and *p*-amino-phenol in wastewater. The optimized conditions were 100 g/L concentration, 40 min contact time, 11 pH, 5 mg/mL nanoparticles amounts, 60 min desorption time, 9 desorption pH and 298 K temperature. Capillary electrophoresis conditions were phosphate buffer (15 mM, pH 7.0) background electrolyte, 18 kV applied voltage, 214 nm UV detection, 30 s sample loading at 23 ± 1 °C. The maximum percent uptakes of *p*-amino-phenol and phenol were 80.0 and 85.0%. High ratio recoveries of *p*-amino-phenol and phenol from nanomaterial were 99.0 and 98. Consequently, the actual extractions of *p*-amino-phenol and phenol from wastewater were 79.2 and 83.30 percent. The migration times of phenol and *p*-amino-phenol and were 9.0 and 12.0 min. The detection limits of phenol and *p*-amino-phenols were 0.1 and 0.2 μg/L after extraction and CE. Therefore, this combination of solid phase micro membrane tip extraction and capillary electrophoresis may be considered as the ideal one for monitoring of toxic phenol and *p*-amino-phenol in water sample.

Keywords: phenol; 3-aminophenol; wastewater; SPMMTE; capillary electrophoresis

1. Introduction

The analysis of any sample of natural origin comprises two parts, that is, sample preparation and separation and identification. The first part of the experiment becomes important in case of the samples of biological and environmental origins. This is due to the fact that, sometimes, more than hundreds of impurities are present in the samples of the natural origin [1,2]. Hence, sample preparation is one of the crucial issues in the separation science. Many sample preparation methods are reported in the literature and these comprise liquid-liquid extraction, solid phase extraction, solid phase micro-extraction (SPME), supercritical fluid extraction, ultrasonic extraction, pressurized liquid extraction, empty fiber liquid phase micro-extraction, micro-wave assisted extraction, dispersive liquid-liquid micro-extraction, molecularly imprinted solid phase extraction, pressurized hot water extraction and solid phase micro membrane tip extraction (SPMMTE). Among these, the later method is supposed as the best one due to the involvement of low amount of adsorption media, fast extraction, capable to work at low concentration and low sample volume. This method is used for the extraction of some species from samples of biological and environmental importance [3,4]. Phenols are very notorious water pollutants and need to be monitored even at low concentration for which SPMMTE may be a choice of the sample preparation. Phenols have acute toxicities [5–15] along with tumorigenesis [16–20]. The collective causes of phenols pollution are industries related to dyes, textiles, pesticides, paper, pharmaceutical, tanning, plastic, gasoline, rubber and so forth [21,22]. Some systems are described for analyses but many of them consume high volume of toxic solvent and sorbent materials [23]. Besides, some techniques are not skilled for phenolic extraction at micro level concentration in water [24]. Taking these assessments into deliberation, composite iron nanomaterial is manufactured and used in SPMMTE for sample preparation of phenols containing water. Analysis of phenol was carried out by capillary electrophoresis [25].

2. Results and Discussion

The characterization of the synthesized composite iron nanoparticles, determination of phenols by capillary electrophoresis and optimization of the extraction of phenols in SPMMTE are discussed.

2.1. Characterization of the Synthesized Composite Iron NanoParticles

The prepared composite iron nanoparticles were categorized by XRD, TEM and SEM methods. The detail of the description is described elsewhere [26–29]. XRD peaks were observed at 2θ of 18.6 and 44.8; indicating presence of FeOOH and Fe^0 (zero valent iron). Further XRD peak show amorphous nature of nanocomposite. TEM results reflect the roughness of the surface and SEM results indicates 10–30 nm particle size.

The supra molecular structure of the composite iron nanoparticle is revealed in Figure 1.

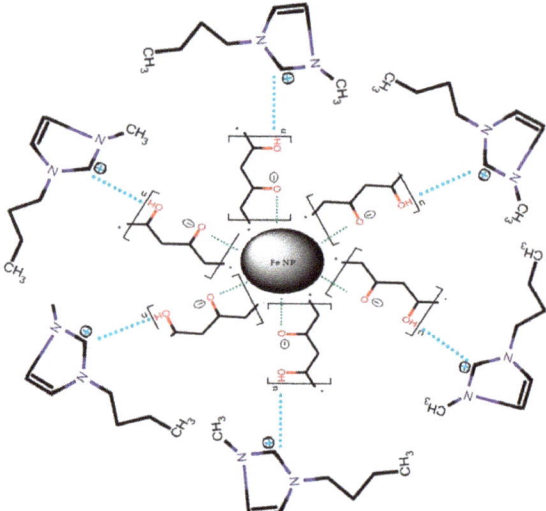

Figure 1. The supra molecular structure of the composite iron nanoparticle.

2.2. Determination of Phenols by Capillary Electrophoresis

Analysis of phenols in water was done by noting their migration times. The documentation of the separated phenols was carried out using standards of the phenols. The values of the migration times of phenol and 3-aminophenol in CE were 9.0 and 12.0 min. The capillary electrophoresis optimization was carried out by achieving the variation in the configuration of background electrolyte, pH of the background electrolyte, amount of sample loaded and capillary length and voltage. The best conditions were used in this paper. The validation data was also collected for this set of experiment and found ±0.70 to ±0.8 as standard deviation, 0.9997 to 0.9998 correlation coefficients and 98.8 to 99.0 confidence level.

2.3. Validation of Capillary Electrophoresis

The capillary electrophoretic method was validated as per standard procedure [30,31]. The valuables studied were linearity, specificity, limit of detection (LOD), limit of quantification (LOQ), accuracy, precision and ruggedness. The magnitudes of these variables are given in Table 1. It is lucid for the table that the developed method is accurate, specific, précised and rugged.

Table 1. Validation parameters of capillary electrophoresis.

Sl. No.	Validated Parameters	% RSD	Correlation Coefficients	Confidence Levels
1.	Linearity	0.55–1.20	0.9996–0.9997	97.0–96.2
2.	LOD	0.85–1.30	0.9995–0.9996	95.5–96.1
3.	LOQ	0.83–0.102	0.9995–0.9996	96.6–96.0
4.	Specificity	0.73–0.93	0.9995–0.9996	97.0–96.1
5.	Accuracy	0.69–0.87	0.9996–0.9998	96.2–96.2
5.	Precision	0.58–0.71	0.9996–0.9998	96.2–97.0
6.	Ruggedness	0.89–1.40	0.9995–0.9996	96.7–96.6

2.4. Extraction of Phenols by SPMMTE

To attain the maximal extraction of phenols with SPMMTE, numerous parameters were diverse and improved. The diverse settings were concentrations of phenols, pH, contact time, nanoparticles amount bounded in a cone of membrane and the temperature.

2.4.1. Concentrations of Phenols

Primarily, with an augment in concentration from 25.0 to 100.0 µg L^{-1} the quantity of sorption in aminophenol and phenol progressively augmented from 5.0 to 16.0 µg/mg that is, 100–80%. Subsequently, more increase in concentration up to 150.0 µg L^{-1}, the sorption of both remained persistent. Extra limitations; interaction time 40 min, pH 11.0, dose 5.0 mg/10 mL and temperature 298 K were attuned. The upshot of this process is given in Figure 2, that evidently displays the sorption of 5.0, 10.0, 13.50, 16.0, 16.0 and 16.0 µg/mg at 25.0, 50.0, 75.0, 100.0, 125.0 and 150.0 µg/L concentrations for amino-phenol. Whereas in the event of phenol sorption of 5.0, 10.0, 14.0, 17.0, 17.0 and 17.0 µg/mg at 25.0, 50.0, 75.0, 100.0, 125.0 and 150.0 µg/L concentrations were gotten. It obviously confirms that an extra increase in concentration from 100.0–150.0 µg/L could not boost the sorption. So, a typical of concentration of 100 µg L^{-1} was preferred for the remaining experiments.

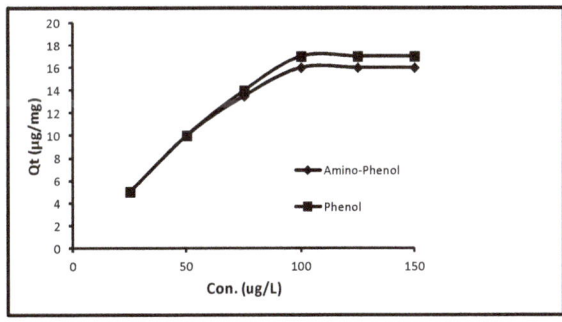

Figure 2. Effect of initial concentrations of phenols.

2.4.2. Extraction Time

With an augment of interaction time from 5.0 to 60.0 min, the sorption frequently gets augmented for aminophenol and phenol till 40.0 min. Then the sorption remained static for both the molecules. Limitations such as concentration, pH, dose and temperature were 100 µg L^{-1}, 11.0, 5.0 mg/10 mL and 298 K were utilized, correspondingly. The contact time effect on sorption is revealed in Figure 3. It may be observed from the figure the sorption in aminophenol and phenol were 4.0, 6.0, 10.0, 14.0 and 16.0 µg/mg that is, 20.0, 30.0, 50.0, 70.0 and 80.0% at 5.0, 10.0, 20.0, 30.0 and 40.0 min and 4.0, 6, 10.50, 14.50 and 17 µg/mg that is, 20.0, 30.0, 50.0, 72.50 and 85.0% at 5.0, 10.0, 20.0, 30.0 and 40.0 min, correspondingly. Additionally, with a rise in contact time the sorption remained unaltered for both molecules. So, at 40.0 min contact time, the rest runs were done.

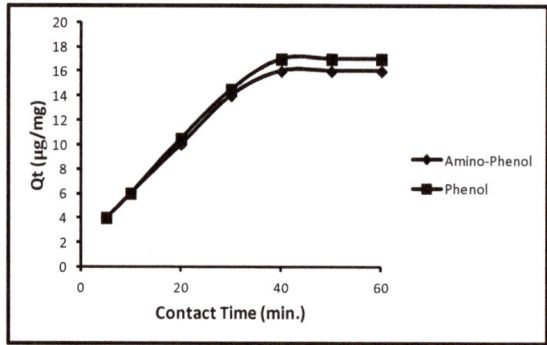

Figure 3. Effect of extraction time of phenols.

2.4.3. pH of the Solutions of Phenols

In the event of pH effect on the sorption, primarily the sorption of aminophenol and phenol increased from 2.0 to 16.0 µg/mg, that is, 10.0–80.0% at 9.0–10.50 pH; followed by the unaffected until 13.0. The extra limitations used were contact time (40.0 min), concentration (100 µg L^{-1}), dose (5.0 mg/10 mL) and temperature (298 K). The effect of pH on sorption is revealed in Figure 4. A review of this figure shows that the quantity of sorption for aminophenol was 2.0, 3.0, 5.0, 16.0 and 16.0 µg/mg, that is, 10.0, 15.0, 25.0, 80.0 and 80.0% at 9.0, 9.50, 10.0, 10.50 and 11.0 pH. Whilst for phenol at similar pH values, the attained sorption percentages were 10.0, 15.0, 25.0, 80.0 and 80.0. Also, for both molecules the further alteration in pH could not augment the sorption. So, a pH of 11.0 remained a standard one. This performance of phenol sorption was because of pKa values of phenol and p-amino-phenol (9.95 and 10.30).

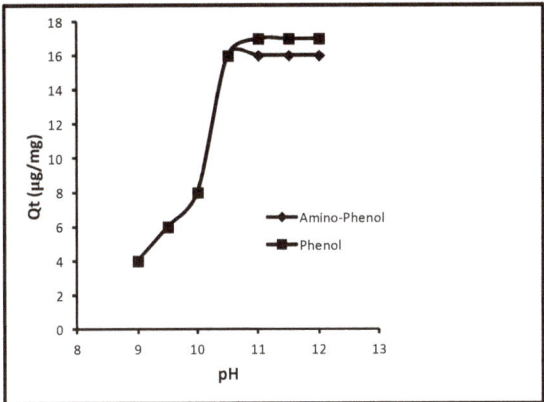

Figure 4. Effect of pH of the solutions of phenols.

2.4.4. Amount of Nanosorbent Enclosed in the Membrane Cone

The effect of the amount of nanosorbent enclosed in the membrane cone was observed while making the other limitations at contact time (40.0 min), pH (11.0), concentration (100 µg/L) and temperature (298 K). The amount of the nanosorbent effect enclosed in the membrane cone on the sorption of phenols is revealed in Figure 5. It was seen that up to 5.0 mg/10 mL, the quantity of sorption gradually augmented and then it remained unchanged. At doses of 1.0, 2.0, 3.0, 4.0, 5.0, 6.0 and 7.0 mg/10.0 mL the removal of 3.0, 7.0, 10.0, 13.0, 16.0, 16.30, 16.50 µg/mg with 15.0, 35.0, 50.0, 65.0, 80.0, 81.50 and 81.50%, correspondingly, was noted in case of aminophenol. Whilst a similar quantity of uptake was 3.0, 7.0, 10.50, 13.50, 17.0, 17.50 and 17.60 µg/mg with 15.0, 35.0, 52.50, 67.50, 85.0, 87.0 and 87.0% for phenol. So, 5.0 mg/mL was the best quantity of nanosorbent enclosed in the membrane cone.

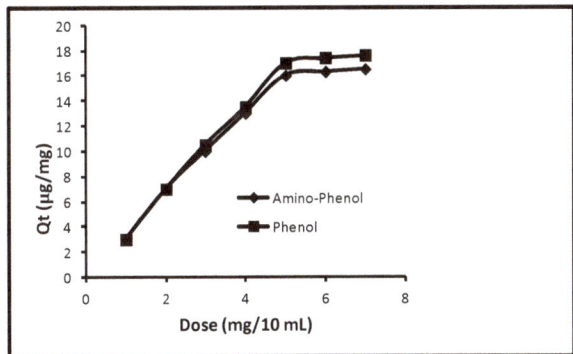

Figure 5. Effect of amount of nanoparticles enclosed in the membrane cone.

2.4.5. Desorption Time

While keeping the parameters such as the 40 min contact time, 100 µg/L concentration, 5.0 mg/10 mL dose and 298 K temperature, the effect of time on desorption for aminophenol and phenol was experimented and it was found that with an increase in time interval from 5 to 60 min, % recoveries of both increased constantly (Figure 6) but, on reaching 60 min, % recoveries remained unaffected that is, constant % recoveries. From Figure 3, it is clear at 5, 10, 20, 30, 40 and 50 min that the % recoveries for aminophenol were 42, 58, 65, 75, 85 and 100 and for phenol 40, 55, 60, 70, 80 and 100, respectively.

Therefore, it is concluded that, with further increase in time interval after 60 min, the desorption could not be changed.

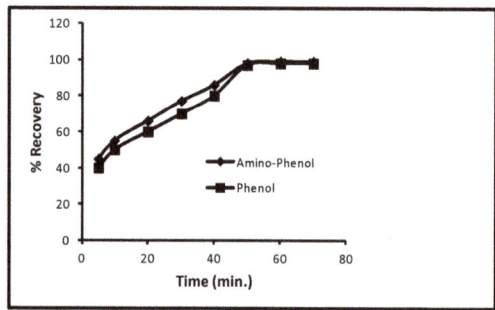

Figure 6. Effect of desorption time for phenols.

2.4.6. Desorption pH

Another experiment in which the effect of pH on the desorption of aminophenol and phenol was tested and other limitations used were a 40 min contact time, 100 µg/L concentration, 5.0 mg/10 mL dose and 298 K temperature. The readings showed that when pH was increased from 9 to 12 for aminophenol and phenol the rate of desorption decreased progressively. The effect of pH on desorption is shown in Figure 7 which depicts, at 9, 9.5, 10, 10.5, 11, 11.5 and 12 pH, that the percentage recovery for aminophenol and phenol was 100, 100, 89, 78, 76, 65 and 45 and 100, 100, 90, 77, 75, 60 and 40 respectively. This performance of phenol sorption was because of the pKa values of phenol and *p*-amino-phenol (9.95 and 10.30).

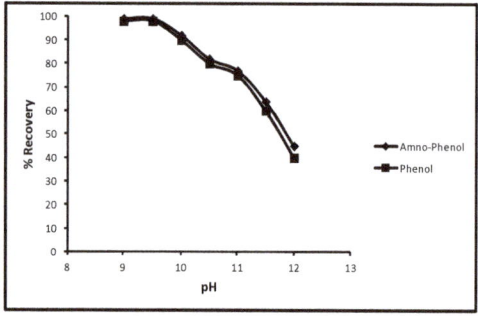

Figure 7. Effect of pH for desorption of phenols.

3. Materials and Methods

3.1. Chemicals and Reagents

Phenol and 3-aminophenol (Figure 8) and nitro-phenol (internal standard) were obtained from Aldrich Chemical Co., St. Louis, MO, USA. Sodium phosphate and sodium dihydrogen phosphate were supplied by SRL, Mumbai, India. *N*-methyl butyl imidazolium bromide, that is, ionic liquid (IL) was obtained from Fluka, Mumbai, India. Ferrous sulphate and poly vinyl alcohol (PVA) were supplied by Qualigens Mumbai, India. Millipore water was collected by a Millipore-Q system made by Bedford, MA, USA. The polypropylene membrane (0.01 µm) was provided by GVS Filtration Technology, Italy. Phosphate buffer (15 mM (pH 7.0)) was arranged utilizing standard protocol. The solutions (10.0–100.0 µg/L) of separable and combination of phenols were arranged in 15.0 milli mole phosphate buffer (pH 7.0) for capillary electrophoresis experiments. The solutions of 0.25–1.5 mg/L were prepared in Millipore water for sample preparation studies.

Phenol.

p-Amino-phenol.

Figure 8. Chemical structures of phenol and *p*-nitro-phenol.

3.2. Instruments Used

The capillary electrophoresis machine was a Quanta 4000 of Waters Chromatography, Millipore, Milord, MA, USA. The software was Millennium 2000 with a data station. The analysis was performed on a fused silica capillary (0.6 m × 75 μm I.D.) and was supplied by Waters, Milord, MA, USA. The pH was adjusted by a pH meter (611, Orion Research Inc., Jacksonville, FL, USA).

3.3. Green Synthesis of Nanomaterial

The nanomaterial, that is, composite iron nanoparticles were manufactured utilizing green knowledge as defined elsewhere [32–35]. Black tea extract was mixed with ferrous sulphate solution and nanoparticles were prepared by green technology. Further, these nanoparticles were allowed to react with ionic liquids to make nanocomposite materials. The instinctive iron nanoparticles were obtained by mingling black tea extract and ferrous sulphate solution. The composite iron nanoparticles were manufactured utilizing *N*-methyl butyl imidazolium bromide ionic liquid.

3.4. Fabrication of SPMMTE

The SPMMTE unit was laboratory made as shown in Figure 9. Poly propylene membrane was molded into a cone shape [35 mm × 35 m × 25 m]. The synthesized nanocomposite (5.0 mg) was inserted into a cone shaped poly propylene membrane. This cone was fixed inside 200.0 μL pipette micro tip. The micro tip was dipped in acetone for about 15 min for fixing.

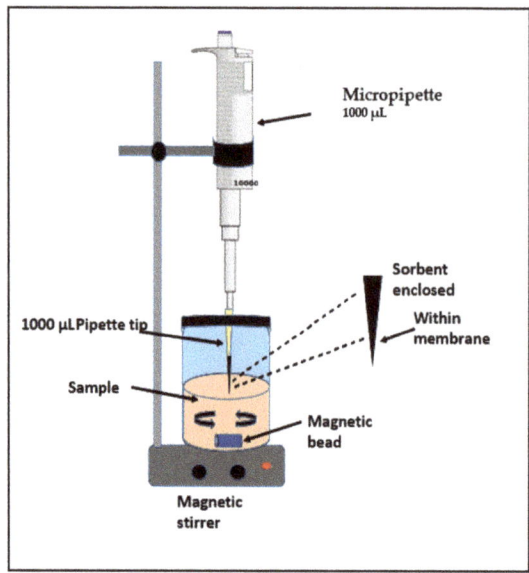

Figure 9. Solid phase micro membrane tip extraction assembly.

3.5. Capillary Electrophoretic Conditions

The capillary electrophoresis machine is described as above. The UV detector (214 nm) was on the cathode side of the machine. The background electrolyte utilized was phosphate buffer (pH 7.0, 15 mM). The experiments were made at 18 kV used voltage. The samples were laden for 30 s via hydrostatic manner of injection. The data were collected at 20 points/second. The experiments were conducted at 23 ± 1 °C temperature. The identification of the separated phenols in the sample was done by comparing the electropherograms of standard ones. The migration times of single phenols were matched with the migration times of the phenols mixture. The quantitative analyses of the phenols in water samples was ascertained using the peak areas of the standard phenols. The internal standard (nitro-phenol) was added before and quantification was achieved by using ratios of peak height or area of the component to the internal standard.

$$\text{Conc.}_{unknown} = [AISK/AISU] \times [AU/AK] \times [\text{Conc.}_{known}], \tag{1}$$

where AISK = area of peak of internal standard in known, AISU = area of peak of internal standard in unknown, AU = area of unknown, AK = area of known.

3.6. Extraction of Phenols from Wastewater by SPMMTE

The solution (0.25–1.5 mg/mL) of the two phenols mixtures were made in pure water. Wastewater samples were collected from the mess pipe line and filtered via Whatman filter paper No. 24 (pre-saturated with phenols). The solutions of the phenol mixture, that is, 1.0 mL of 0.25–1.5 mg/L concentrations were mixed to 9.0 mL water samples to get concentrations from 25–150 µg/L. The wastewater sample was shaken for about 60 min and kept at room temperature overnight. The extraction of phenols from wastewater samples was carried out using the SPMMTE method.

The SPMMTE tip was dipped in the wastewater sample (10 mL) with constant stimulation with a magnetic stirrer for 30 min. A 200 µL sample was engulfed in the SPMMTE tip at a gap of 15 s. The engulfed wastewater was freed back into the beaker. This exercise was performed for 40 min. The cone shaped membrane was taken away from SPMMTE and air dried. This cone shaped structure was shaken with methanol (25 mL) to extract the phenols. The methanol solution was condensed to 0.5 mL in a vacuum. The resultant solution was utilized for determining the concentrations of the phenols using capillary electrophoresis.

4. Conclusions

From the results and discussion, it is clear that the combination of solid phase micro membrane tip extraction and capillary electrophoresis is ideal for the monitoring of toxic phenol and *p*-amino-phenol in wastewater. The detection limits of phenol and *p*-amino-phenols were 0.1 and 0.2 µg/L after extraction and CE. The percentage recoveries are quite satisfactory for both the phenols. The method is also fast as it can be finished within 12 min. The optimized circumstances were 100 g/L concentration, 40 min contact time, 11 pH, 5 mg/mL nanoparticles amounts, 60 min desorption time, 9 desorption pH and 298 K temperature. The maximum percent uptakes of *p*-amino-phenol and phenol were 80.0 and 85.0%. The maximum percentage recoveries of *p*-amino-phenol and phenol from the nanomaterial were 99.0 and 98 %. Consequently, the actual extractions of *p*-amino-phenol and phenol from wastewater were 79.2 and 83.30 percent. Therefore, this combination of solid phase micro membrane tip extraction and capillary electrophoresis may be considered as suitable for the monitoring of toxic phenol and *p*-amino-phenol in any water sample.

Author Contributions: Data curation, N.H.A.-S.; Formal analysis, Z.A.A.; Methodology, L.H.A-W.; Software, H.A.; Supervision and paper writing, I.A.

Funding: This work was funded by the Deanship of Scientific Research at Princess Nourah bint Abdulrahman University, through the Research Groups Program Grant no. (RGP-1440-0017).

Conflicts of Interest: There is no conflict of interest.

References

1. Ali, I.; Gupta, V.K.; Aboul-Enein, H.Y. Hyphenation in sample preparation: Advancement from the micro to the nanoworld. *J. Sep. Sci.* **2008**, *31*, 2040–2053. [CrossRef] [PubMed]
2. Ali, I.; Kulsum, U.; ALOthman, Z.A. Advances in analyses of profens in biological and environmental samples by liquid chromatography. *Curr. Pharm. Anal.* **2016**, *12*, 158–176. [CrossRef]
3. Ali, I.; Abbasi, J.; Alothman, Z.A.; Alwarthan, A. SPMMTE and HPLC methods for the analyses of cardiovascular drugs in human plasma using new generation C_{28} column. *Cur. Pharm. Anal.* **2016**, *13*, 56–62. [CrossRef]
4. Ali, I.; Rani, D.; Alothman, Z.A. Analysis of ibuprofen, pantoprazole and itopride combination therapeutic drugs in human plasma by solid phase membrane micro tip extraction (SPMMTE) and high performance liquid chromatography (HPLC) methods using new generation Core Shell C_{18} column. *J. Liq. Chromatogr. Rel. Technol.* **2016**, *39*, 339–345. [CrossRef]
5. *The Merck Index: An Encyclopedia of Chemical, Drugs and Biological*; Budavari, S.S. (Ed.) RSC Publishing: London, UK, 2013.
6. Dean-Ross, D.; Rahimi, M. Toxicity of phenolic compounds to sediment bacteria. *Bull. Environ. Contam. Toxicol.* **1995**, *55*, 245–250. [CrossRef] [PubMed]
7. Basheer, A.A. Chemical chiral pollution: Impact on the society and science and need of the regulations in the 21st century. *Chirality* **2018**, *30*, 402–406. [CrossRef] [PubMed]
8. Basheer, A.A. New generation nano-adsorbents for the removal of emerging contaminants in water. *J. Mol. Liq.* **2018**, *261*, 583–593. [CrossRef]
9. Basheer, A.A.; Ali, I. Stereoselective uptake and degradation of (±)-o, *p*-DDD pesticide stereomers in water-sediment system. *Chirality* **2018**, *30*, 1088–1095. [CrossRef]
10. Ali, I.; Basheer, A.A.; Mbianda, X.Y.; Burakov, A.; Galunin, E.; Burakova, I.; Mkrtchyan, E.; Tkachev, A.; Grachev, V. Graphene based adsorbents for remediation of noxious pollutants from wastewater. *Environ. Int.* **2019**, *127*, 160–180.

11. Ali, I.; Alharbi, O.M.L.; Alothman, Z.A.; Badjah, A.Y.; Abdulrahman, A. Artificial neural network modelling of amido black dye sorption on iron composite nanomaterial: Kinetics and thermodynamics studies. *J. Mol. Liq.* **2018**, *250*, 1–8. [CrossRef]
12. Ali, I.; Al-Othman, Z.A.; Alwarthan, A. Removal of secbumeton herbicide from water on composite nanoadsorbent. *Desal. Water Treat.* **2016**, *57*, 10409–10421. [CrossRef]
13. Ali, I.; Asim, M.; Khan, T.A. Arsenite removal from water by electro-coagulation on zinc–zinc and copper–copper electrodes. *Int. J. Environ. Sci. Technol.* **2013**, *10*, 377–384. [CrossRef]
14. Ali, I.; Gupta, V.K.; Aboul-Enein, H.Y. Metal ion speciation and capillary electrophoresis: Application in the new millennium. *Electrophoresis* **2005**, *26*, 3988–4002. [CrossRef] [PubMed]
15. Ali, I. Microwave assisted economic synthesis of multi walled carbon nanotubes for arsenic species removal in water: Batch and column operations. *J. Mol. Liq.* **2018**, *271*, 677–682. [CrossRef]
16. DellaGreca, M.; Monaco, P.; Pinto, G.; Pollio, A.; Previtera, L.; Temussi, F. Phytotoxicity of low-molecular-weight phenols from olive mill waste waters. *Bull. Environ. Contam. Toxicol.* **2001**, *67*, 352–359. [CrossRef]
17. Hirose, M.; Takesada, Y.; Tanaka, H.; Tamano, S.; Kato, T.; Shirai, T. Carcinogenicity of antioxidants BHA, caffeic acid, sesamol, 4-methoxyphenol and catechol at low doses, either alone or in combination and modulation of their effects in a rat medium-term multi-organ carcinogenesis model. *Carcinogenesis* **1998**, *19*, 207–212. [CrossRef] [PubMed]
18. Hooived, M.; Heederik, D.J.; Kogevinas, M.; Boffetta, P.; Needham, L.L.; Patterson, D.G., Jr.; Bueno-de-Mesquita, H.B. Second follow-up of a Dutch cohort occupationally exposed to phenoxy herbicides, chlorophenols and contaminants. *Am. J. Epidemiol.* **1998**, *147*, 891–901. [CrossRef]
19. Michałowicz, J.; Duda, W. Phenols–Sources and Toxicity. *Pol. J. Environ. Stud.* **2007**, *16*, 347–362.
20. Schweigert, N.; Zehnder, A.J.B.; Eggen, R.I.L. Chemical properties of catechols and their molecular modes of toxic action in cells, from microorganisms to mammals: Minireview. *Environ. Microbiol.* **2001**, *3*, 81–91. [CrossRef]
21. Ali, I.; Aboul-Enein, H.Y. *Chiral Pollutants: Distribution, Toxicity and Analysis by Chromatography and Capillary Electrophoresis*; John Wiley & Sons: Chichester, UK, 2004.
22. Ali, I.; Aboul-Enein, H.Y.; Gupta, V.K. *Nanochromatography and Capillary Electrophoresis: Pharmaceutical and Environmental Analyses*; Wiley & Sons: Hoboken, NJ, USA, 2009.
23. Ali, I.; Gupta, V.K.; Saini, V.K.; Aboul-Enein, H.Y. Analysis of phenols in wastewater using capillary electrophoresis and solid phase extraction. *Int. J. Environ. Pollut.* **2006**, *27*, 95–103. [CrossRef]
24. Ali, I.; Aboul-Enein, H.Y. Fast Screening of phenol and its derivatives in wastewater by HPLC by using monolithic silica column and solid phase extraction. *Anal. Lett.* **2004**, *37*, 235–2361. [CrossRef]
25. Ali, I.; Aboul-Enein, H.Y. Determination of phenol and its derivatives in waste water by capillary electrophoresis. *Fresenius Environ. Bull.* **2002**, *11*, 36–39.
26. Ali, I.; ALOthman, Z.A.; Sanagi, M.M. Green synthesis of iron nano-impregnated adsorbent for fast removal of fluoride from water. *J. Mol. Liq.* **2015**, *211*, 457–465. [CrossRef]
27. Ali, I.; Al-Othman, Z.A.; Alharbi, O.M.L. Uptake of pantoprazole drug residue from water using novel synthesized composite iron nanoadsorbent. *J. Mol. Liq.* **2016**, *218*, 465–472. [CrossRef]
28. Ali, I.; Kulsum, U.; ALOthman, Z.A.; Saleem, K. Analyses of nonsteroidal anti-inflammatory drugs in human plasma using dispersive nanosolid-phase extraction and high-performance liquid chromatography. *Chromatographia* **2016**, *79*, 145–157. [CrossRef]
29. Ali, I.; ALOthman, Z.A.; Alwarthan, A. Sorption, kinetics and thermodynamics studies of atrazine herbicide removal from water using iron nano-composite material. *Int. J. Environ. Sci. Technol.* **2016**, *13*, 733–742. [CrossRef]
30. United States Pharmacopeial Convention Inc. *United States Pharmacopeia*; United States Pharmacopeial Convention Inc.: Rockville, MD, USA, 2005.
31. IFPMA. *ICH Draft Guidelines on Validation of Analytical Procedures: Definitions and Terminology*; Federal Register; IFPMA: Geneva, Switzerland, 1995; Volume 60.
32. Ali, I.; ALOthman, Z.A.; Alwarthan, A. Molecular uptake of congo red dye from water on iron composite nanoparticles. *J. Mol. Liq.* **2016**, *224*, 171–176. [CrossRef]

33. Ali, I.; Kulsum, U.; ALOthman, Z.A.; Alwarthan, A.; Saleem, K. Functionalized nanoparticles based solid-phase membrane micro-tip extraction and high-performance liquid chromatography analyses of vitamin B complex in human plasma. *J. Sep. Sci.* **2016**, *39*, 2678–2688. [CrossRef]
34. Ali, I.; ALOthman, Z.A.; Alwarthan, A. Green synthesis of functionalized iron nanoparticles and molecular liquid phase adsorption of ametryn from water. *J. Mol. Liq.* **2016**, *221*, 1168–1174. [CrossRef]
35. Ali, I.; ALOthman, Z.A.; Alwarthan, A. Synthesis of composite iron nanoadsorbent and removal of ibuprofen drug residue from water. *J. Mol. Liq.* **2016**, *219*, 858–864. [CrossRef]

Sample Availability: Samples of the compounds are available from the authors.

© 2019 by the authors. Licensee MDPI, Basel, Switzerland. This article is an open access article distributed under the terms and conditions of the Creative Commons Attribution (CC BY) license (http://creativecommons.org/licenses/by/4.0/).

Article

Heavy Metals Bioindication Potential of the Common Weeds *Senecio vulgaris* L., *Polygonum aviculare* L. and *Poa annua* L.

Mirko Salinitro [1], Annalisa Tassoni [1], Sonia Casolari [2], Francesco de Laurentiis [2], Alessandro Zappi [2] and Dora Melucci [2,*]

[1] Department of Biological Geological and Environmental Sciences, University of Bologna, Via Irnerio 42, 40126 Bologna, Italy
[2] Department of Chemistry "G. Ciamician", University of Bologna, Via Selmi 2, 40126 Bologna, Italy
* Correspondence: dora.melucci@unibo.it; Tel.: +39-051-2099530; Fax: +39-051-2099456

Received: 15 July 2019; Accepted: 31 July 2019; Published: 1 August 2019

Abstract: In recent years, heavy metals (HMs) levels in soil and vegetation have increased considerably due to traffic pollution. These pollutants can be taken up from the soil through the root system. The ability of plants to accumulate HMs into their tissues can therefore be used to monitor soil pollution. The aim of this study was to test the ruderal species *Senecio vulgaris* L., *Polygonum aviculare* L., and *Poa annua* L., as possible candidates for biomonitoring Cu, Zn, Cd, Cr, Ni and Pb in multiple environments. The soils analyzed in this work came from three different environments (urban, woodland, and ultramafic), and therefore deeply differed for their metal content, texture, pH, and organic matter (OM) content. All urban soils were characterized by high OM content and presence of anthropogenic metals like Pb, Zn, Cd, and Cu. Woodland soils were sandy and characterized by low metal content and low OM content, and ultramafic soils had high Ni and Cr content. This soil variability affected the bioindication properties of the three studied species, leading to the exclusion of most metals (Zn, Cu, Cr, Cd, and Pb) and one species (*P. aviculare*) due to the lack of linear relations between metal in soil and metal in plants. *Senecio vulgaris* and *Poa annua*, conversely, appeared to be good indicators of Ni in all the soils tested. A high linear correlation between total Ni in soil and Ni concentration in *P. annua* shoots ($R^2 = 0.78$) was found and similar results were achieved for *S. vulgaris* ($R^2 = 0.88$).

Keywords: bioindication; heavy metals; urban soil; *Senecio vulgaris*; Poa annua; Polygonum aviculare; predictive models

1. Introduction

The increasing urbanization and industrialization in the last decades has resulted in spreading of heavy metals in the environment [1–3]. Since these elements are not degradable, they slowly accumulate in soil, becoming potentially hazardous to terrestrial and aquatic ecosystems and a risk to human and animal life [4,5]. Heavy metals (HMs) are naturally present in the environment as a result of either natural processes and human activities [6,7]. In natural ecosystems, HMs come from ultramafic rocks and ore minerals, and during weathering that leads to soil formation they can be released in the environment [8]. On the other hand, anthropogenic sources (i.e., vehicular traffic, mining activities, and refining processes) are nowadays the main responsible for HMs pollution [9,10]. Urban areas are recognized to be major sources for contaminants [11,12], and traffic is the primary source of HMs that accumulate in roadside soils [13] and street dust [14]. HMs are produced by vehicles exhaust emissions, as well as from the wear and tear of mechanical parts such as brakes, tires, and catalytic converters [13,15].

In recent years, it has been demonstrated that HMs levels in soil and vegetation have increased considerably because of traffic pollution, and the problem keeps growing with the increase of vehicular traffic [16]. This diffuse source of pollution in areas where people live and food is produced pose a serious threat to human health. In fact, these pollutants can enter plants directly via foliar absorption or be taken up from the soil through the root system [17] and undergo processes of biomagnification [18,19]. Despite the high toxicity of HMs for plants, when these elements are present in the soil at low concentrations, plants continue to grow healthy even while accumulating these metals. The ability of plants to accumulate HMs into their tissues can be used as monitoring tool to asses soil pollution [20], even though it has to be taken into account that many factors (i.e., climate, metal availability, etc.) could influence HMs uptake by plants.

Biomonitoring techniques using indicator plants (bioindication) are becoming common methods to detect toxic levels of HMs in the environments. Mosses and lichens, for example, are known to be the most sensitive indicators of atmospheric pollution [21,22], thus they are broadly used in urban environment. Unfortunately, because of the absence of roots and their restricted presence on hard substrate, they are not suitable for soil monitoring. Many authors agree that herbaceous plants could be effective biomonitoring tools, and some common species like the Dandelion, Nettle, and Broadleaf Plantain have already been successfully used [20,23–25]. Not all plants are suitable as indicators; some basic characteristics of good bioindicators have been listed by Witting [26]. An indicator plant should (i) accumulate one or several selected elements, (ii) have low sensitivity to the accumulated elements, (iii) have wide distribution in various environments, and (iv) show a correlation between metal accumulation and input into the ecosystem.

Even when using the right plant, bioindication properties could be affected by other factors like soil properties complexation of HMs, oxidation state [8,27] and phenologic phase of the plants [28]. Seasonality plays an important role in determining HMs concentration in plant tissues. Despite some species grow all year long, it has been demonstrated that Cu, Fe, Mn, Pb, and Zn contents in Dandelion leaves collected in autumn were higher compared to those collected at the same sites in the spring [28]. Similar results were reported for alfalfa with regard to Mo content [29]. Soil properties, like cationic exchange capacity, clay content, pH, and organic matter content, are likely to change HMs availability to plants [30,31]. For example, Dai et al. [32] estimated that extractable Cd, Pb, and Zn levels in contaminated soils were positively correlated with organic matter contents. Moreover, low pH is optimal for metal availability since solubility has been shown to increase with decreasing pH [33,34]. Given the high variability of soils, bioindication cannot be considered a technique to precisely measure trace metals in soil, but rather a way to estimate them and their interaction with plants in some specific conditions. Therefore, it is of vital importance to assess indication properties of a species on several soils that differ for their HMs content and physical properties.

In this study we did not take into account the different forms of metals present in soils —e.g., Cr(III)/Cr(VI)— and metal complexes (e.g., Pb/tetraethyl-Pb) since we did not aim at the evaluation of the hazard of the different HMs and their forms. The aim of our study was to evaluate the ruderal species *Senecio vulgaris* L., *Polygonum aviculare* L., and *Poa annua* L. as possible candidates for biomonitoring Cu, Zn, Cd, Cr, Ni, and Pb in multiple environments, with special attention to the urban ones. These species have never been tested for biomonitoring potential, but since their wide diffusion in all anthropic environment, they could help in detailed biomonitoring campaigns in urban areas. Furthermore, we aimed to assess how different types of soils can affect the predictive potential of these species.

2. Materials and Methods

2.1. Samples Collection

For this study, three common weed species that have the basic characteristics of "good bioindicators" according to Witting [26] were selected. All three species are ruderal plants, which makes them common in all anthropic habitats.

(1) *Senecio vulgaris* L. (groundsel) is an annual plant of the Asteraceae family. Originally having an Eurasiatic distribution, today it has become subcosmopolitan worldwide. The species is common everywhere, and it grows prolifically in disturbed habitats like road margins, arable fields, and gardens. It preferably grows on clayey soil rich in nitrogen and organic matter.
(2) *Poa annua* L. (annual bluegrass) is an annual plant of the Poaceae family. Originally having an Eurasiatic distribution, today it is widely naturalized in the temperate areas of the globe. It is a pioneer species that grows in trampled areas, gardens, and roads margins, in nitrogen-rich soils.
(3) *Polygonum aviculare* L. (common knotgrass) is an annual plant of the Polygonaceae family. The species is cosmopolitan, and because of its high variability it is adaptable to several habitats. It grows on all soil types and it is resistant to trampling. It is widespread in urban areas, arable fields, but also woodland margins.

To maximize the metal content in plant tissues, plants were harvested close to the end of their life cycle, therefore during the fruiting season (April–May 2017 for *P. annua* and *S. vulgaris*; October 2017 for *P. aviculare*). Only the aerial parts of the plants were taken. After collection, plants where thoroughly washed with deionized water, then oven dried at 50 °C until constant weight. Dried samples were powdered with an Ultraturrax A11 basic Analytical mill (IKA®, Staufen, Germany), then stored at room temperature until analysis.

In order to have high heterogeneity of soil conditions, the three herbaceous species were harvested in eight stations belonging to three different habitats (Figure 1). Five stations were from urban environment (BO7, BO8, MI3, MI4, and MI9; "BO" indicating samples collected in Bologna and "MI" collected in Milan), two stations were from woodland environment (NAT1 and NAT5; "NAT" indicating natural environment) and one was an ultramafic station from Mount Prinzera (NAT8). The choice criteria were: most and least polluted stations in urban environments, random choice among woodland station (since they were all similar), and the only ultramafic station.

In every station, the three species were all present simultaneously, growing in the same bulk soil. The soil was sampled at a depth of 0–10 cm exactly below the plants, and at least 5 soil subsamples were collected in each location. The subsamples were then mixed together to form one bulk sample (of ~2 kg). The bulk soil samples were homogenized and sieved at 0.5 cm to exclude stones and other coarse particles, then oven dried at 50 °C until constant weight. Dried soil samples were further sieved at 0.1 cm, then stored at room temperature until analysis.

2.2. Soil Digestion

All chemicals used were suprapure and were purchased from Sigma-Aldrich (Saint Louis, MO, USA): 69% (w/w) HNO_3, HCl 37% (w/w), 35% (w/w) H_2O_2, 96% (w/w) sulfuric acid, ammonium citrate, iron (II) ammonium sulfate hexahydrate, and potassium dichromate. RC syringes filter (Fisher Scientific, Hampton, NH, USA) (porosity 0.45 μm; diameter 22 mm) were employed to filter solutions after metals extraction.

To perform soil digestion, a modified a version of the US EPA 3050b method was used. The dried and sieved soil (0.1 cm) was finely grinded in a mortar; then, approximately 1 g of soil and 5 mL of 69% (w/w) HNO_3 were put in a Pyrex 100-mL calibrated test tube. The tube was connected to a Vigreux column, and was then placed in a special housing on a heating plate at 150 °C. The system was left in reflux mode for 30 min. Subsequently, the tube was cooled in an ice-bath, then 5 mL of 35% (w/w) H_2O_2 was added, and the addition of H_2O_2 drops continued until the solution in the tube stopped boiling. Then, 10 mL of 37% (w/w) HCl were introduced, and another reflux step was applied for 15 min. Once digestion was completed, the solution was cooled down and filtered. Finally, the liquid phase was transferred into a 50-mL flask and brought to the final volume with 0.5 M HNO$_3$. The total concentration of metals in soils (Total Metal, TM) was measured in µg g^{-1} (briefly indicated as "ppm"). Blank digestions (without soil) were carried out using the same reagents as described above.

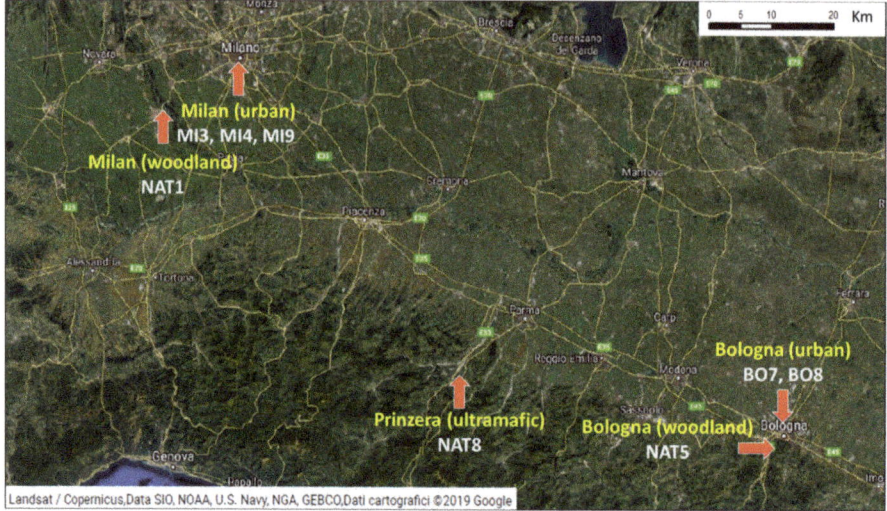

Figure 1. Sampling locations of soils and plants used in the study. In each station one soil sample and three plant species have been collected.

2.3. Plant Digestion

Acid digestion on plant shoots was carried using a modified protocol adapted from Huang et al. [35]. An aliquot of plant powdered shoots, between 0.05 and 0.1 g, was placed in a 10 mL glass tube and 2 mL of 69% (w/w) HNO$_3$ were added. A pre-digestion phase was obtained by leaving the tubes at room temperature for 24 h. Then, the tubes were placed on a hot plate at 75 °C for 1 h, and subsequently the temperature was increased to 125 °C for another 1 h. During the digestion the tubes were left open without any reflux system.

In the 2 h digestion, the volume of acid reduced to about 1 mL, and then it was transferred in a 10 mL flask and brought to the final volume with Milli-Q grade water (Millipore, Bedford, MA, USA) to obtain a digestate with about 6–7% (w/w) HNO$_3$. Generally, no plant residues were visible, but for the sake of sureness, filter syringes were used. The concentration of metals in plants (Plant Metal, PM) was measured in µg g^{-1} (briefly indicated as "ppm"). Blank digestions (without plants) were carried out using the same reagents as described above.

2.4. Extraction of Bioavailable Metal Fraction from Soil

The dried and sieved soil (0.1 cm) was finely ground in a mortar, then sieved with a 0.5 mm mesh. Five grams of sieved soil was then transferred to an extraction bottle in which 5 mL of 2% (w/v) ammonium citrate solution was added. The obtained mixture was shaken on an end-over-end tube roller mixer at 30 rpm for 1 h at 20 °C. The extracts were immediately separated by decantation for few minutes, followed by centrifugation for 10 min at about 3000 Ug. The supernatant was recovered and the liquid was stored in a polyethylene container at 4 °C until analysis. The concentration of bioavailable metals in soil (BM) was measured in µg g^{-1} (briefly indicated as "ppm"). Blank extractions (without soil) was carried out using the same reagents as described above [36].

2.5. Determination of Metal Concentration in Soils and Plants

The concentration of metal in soils and plants was quantified through Atomic Absorption Spectroscopy (AAS). AAS measurements were performed using an Atomic Absorption Spectrometer AAnalyst 400 (Perkin-Elmer, Waltham, MA, USA), equipped with a deuterium background corrector, Autosampler AS-72 (Perkin-Elmer) and a HGA 800 graphite furnace (Perkin-Elmer). Single-element Lumina (Perkin-Elmer) hollow-cathode lamps were used. All measurements were carried out using default program for ashing and atomization curves for each element, at the detailed instrumental conditions are reported in Table 1. All the elements, except Zinc, were determined by electro-thermal AAS (ET-AAS), employing argon at flow rate 250 mL min^{-1} in all steps except during atomization (0 mL min^{-1}).

Table 1. Instrument settings for AAS determination.

Element	Wavelength (nm)	Slit (nm)	Drying Temperature (°C)	Pyrolisis Temperature (°C)	Atomization Temperature (°C)
Zn (II)	213.9	0.70	110	700	1800
Cu (II)	324.8	0.80	110	1000	2300
Pb (II)	283.3	1.05	110	950	1800
Cr (VI)	357.9	0.80	110	1650	2500
Cd (II)	228.8	1.35	110	850	1650
Ni (II)	232.0	1.35	110	1400	2500

Zinc was analyzed by flame atomic absorption (FAAS) employing Acetylene (4.10 L min^{-1}) and air (10 L min^{-1}).

For each of the six analyzed metals (Cu, Pb, Cd, Cr, Ni, and Zn) a calibration line was created. Standards for calibration lines were purchased by Merck (Darmstadt, Germany). Three standard solutions were prepared for each metal and "outer" standard concentrations were selected in order to stay in the linear range of each analyte, as tabulated in the software WinLab 32 (Perkin-Elmer). Peak area was used as the analytical signal, after verifying that peak height never overcame 0.6 AU, in order to stay in the absorbance linear range. Before each analysis, a blank sample was analyzed and the peak area of the sample was subtracted to the previous blank one. For each calibration line, the limit of detection (LoD) was computed by applying the equation LoD = $(K\, s_{y/x})/b$ [37], where $s_{y/x}$ and b are the estimated regression standard deviation and the slope of the relevant analytical calibration function, respectively. $K = 3$ was chosen in order to obtain the limits of detection. It was verified that LoD never overcame the lowest standard concentration. When analyses had to be carried out in several days, every day three standards were analyzed and projected on the calibration line, to verify its validity. Three replicates were analyzed for each standard and sample. The injected volume was 20 µL for each analysis. Samples were properly diluted in order to obtain a signal in the calibration range, and the dilution factor was kept into account to calculate the metal concentration in the sample.

In order to analyze Cd, Cr, Ni, and Pb by AAS, some matrix modifiers were necessary. In particular, $Mg(NO_3)_2$ (Perkin-Elmer) for Cd and Cr, $PdCl_2$ (Fluka, Honeywell, Morris Planes, NJ, USA) for Cd, and $NH_4H_2PO_4$ (Sigma Aldrich) for Pb. A solution (20 µL mL^{-1}) containing all of the modifiers was added to each sample; final concentrations: 200 mg L^{-1} for $Mg(NO_3)_2$ and 2.3 mg L^{-1} for $PdCl_2$, 4 mg L^{-1} for $NH_4H_2PO_4$. It was also verified that the presence of an unnecessary modifiers did not influence the measurements of other metals (as Cu and Zn, which did not require any modifier).

2.6. Determination of Organic Matter and Granulometry of Soil

The percentage of organic matter in soils (OM) was determined by two experimental methods: titration and Loss on Ignition. The percentage of inorganic matter (IM) was calculated as 100%-OM.

To measure OM, the titration was carried out following the method in Walkley [38]. Half a gram of soil, 10 mL of potassium dichromate 0.167 M, and 20 mL of 96% (w/w) sulfuric acid were placed in a 500-mL conical flask, slowly percolating along the internal walls of the flask, not to overheat the mixture. The flask was covered with watch glass and left to rest for 30 min. Then the reaction was interrupted by adding 200 mL of distilled water, previously cooled in the refrigerator. A few drops of ferroin (redox indicator) were added, and titration was carried out with a solution of iron(II) ammonium sulfate hexahydrate 0.5 M until the color changed. At the same time, a blank test was performed with 10 mL of dichromate, 20 mL of sulfuric acid, and 200 mL of distilled water. The following expression was used for the calculation of organic carbon (C) expressed in g kg^{-1}.

$$C = 3.9 \cdot \frac{(B-A)}{MSoil} \cdot MFe \tag{1}$$

where B = volume of the solution of iron (II) ammonium sulfate hexahydrate used in the titration of the blank test, expressed in mL; A = volume of the solution of iron (II) ammonium sulfate hexahydrate used in the titration of the sample solution, expressed in mL; MFe = effective molarity of the solution of iron (II) ammonium sulfate hexahydrate; and MSoil = mass of the soil sample, expressed in grams. To transform g kg^{-1} of organic carbon into the corresponding organic substance content, a conversion factor is applied: % OM_titr = %OM_titr·1.724.

To validate OM content found by titration, we compared the results with the one found by Loss on Ignition method (official method for the determination of OM) as explained in Storer [39]. The two methods were comparable and gave similar results, therefore the titration method results were validated and used for statistical elaborations.

Granulometry of the samples was assessed by sieving samples with gradually smaller mashes and weighing the fraction held into each mesh. Four classes of granulometry were defined: particles >0.5 mm (coarse), particles between 0.5 and 0.25 mm (medium), particles between 0.25 mm and 63 μm (fine), and particles <63 μm (ultrafine).

2.7. Data Analysis

The matrix containing all collected data was composed by 86 observation (objects) and 43 variables.

We used chemometrics to extract useful information from our dataset and in particular to create and validate models. Chemometrics was applied both in univariate mode (analysis of correlation and creation of linear regression) and in multivariate mode (Principal Components Analysis (PCA)) [39,40]. To create linear models, the Multiple Linear Regression tool (MLR) was applied. In order to validate MLR models, besides considering the model p-values (ANOVA test), which should be close to the null value, the same data used to create the model were projected onto it, both in "calibration" mode (blue dots in response plots) and by leave-one-out cross-validation (LOO-CV) (red dots in response plots). Projection in calibration mode means that once the model is created, data are projected onto it as they are. LOO-CV, on the other side, creates as many models as the number of samples, leaving each time one sample out from the model creation and projecting it onto such model. In this way, each sample is treated as if it was an external data used to validate the regression performance of the overall model. Both in calibration and in LOO-CV, response values of each sample are recalculated by projection. Then, two further response lines are computed (blue for calibration mode; red for LOO-CV), in which the independent variables are the known (experimental) response and the dependent variables are the recalculated values that appear in the response plot. The predictive model performances are evaluated by the parameters of these lines. In a perfect case, the recalculated responses would be exactly equal to the known ones, thus the response lines would be the bisector of the response plot, with slope = 1, offset = 0, and R^2 = 1. Model performances are considered acceptable if the response line parameters are close to these ideal values. A further parameter of the response lines is root mean squared error (RMSE), which is a sort of sum of distances between the known responses and the recalculated ones, therefore it should be close to zero.

The potential suitability of a species as bioindicator of one or more metals was validated in two steps. Firstly, we tested the correlation between bioaccumulation factor (BAF) calculated on TM and BAF calculated on BM, if that correlation was good (between 0.7 and 1), we tested the correlation between metal content in plants (PM) and metal content in soil (TM). In Table 2, the empirical rules adopted to evaluate correlation "goodness" between variables are shown. Only plants showing "high" or "excellent" correlation values for both validation steps were used in the creation of predictive models. All statistical analyses and graphical elaborations were performed using the software The Unscrambler 10.4 (CAMO Analytics, Oslo, Norway).

Table 2. Empirical rules adopted in the evaluation of correlations. Colors indicate the "goodness" of correlation: yellow = significant; orange = relevant; green = high; blue = excellent.

0.3<correlation<0.5	significant
0.5<correlation<0.7	relevant
0.7<correlation<0.9	high
0.9<correlation<1	excellent

3. Results

3.1. Heavy Metals in Soil

The eight soils analyzed appeared to be strongly different one the other and characterized by various metal content, texture, pH, and OM content. Soil variability was explored trough a PCA (Figure 2) that showed a clustering of soils according to the area of collection.

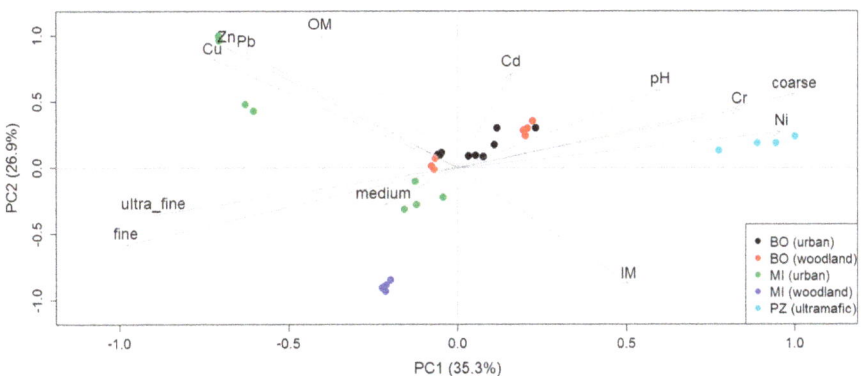

Figure 2. Soil clustering after principal component analysis (PCA). The input data were the soil variables of granulometry, OM, IM, and total heavy metals concentration.

Heavy metals (HMs) concentrations, were closely linked to the levels of anthropogenic activity for urban and woodland soils, while were mainly from geogenic origin in ultramafic soils. The analyzed metals were present in the decreasing order: Zn > Cu > Cr > Ni > Cd for urban areas, Zn > Cu > Ni > Cr > Cd for woodland areas and Ni > Cr > Zn > Cu > Pb > Cd ultramafic areas. All urban soils were characterized by medium to fine granulometry, high OM content especially for Milan samples and presence of anthropogenic metals like Pb, Zn, Cd, and Cu. Woodland soils from Bologna were quite similar to urban ones with a slightly coarser texture and lower pH levels. Milan woodland soils were sandy and characterized by low metal content and low OM content. Finally, Prinzera soils, because of their ultramafic origin, had high Ni and Cr content; moreover, they are characterized by coarser granulometry (Figure 3). Metal concentrations (both total and bioavailable) in each soil are summarized in Table 3. The most polluted soils were MI3 and MI4, while the least polluted were NAT1 and NAT5, except for Ni and Cr where NA8 had the highest values.

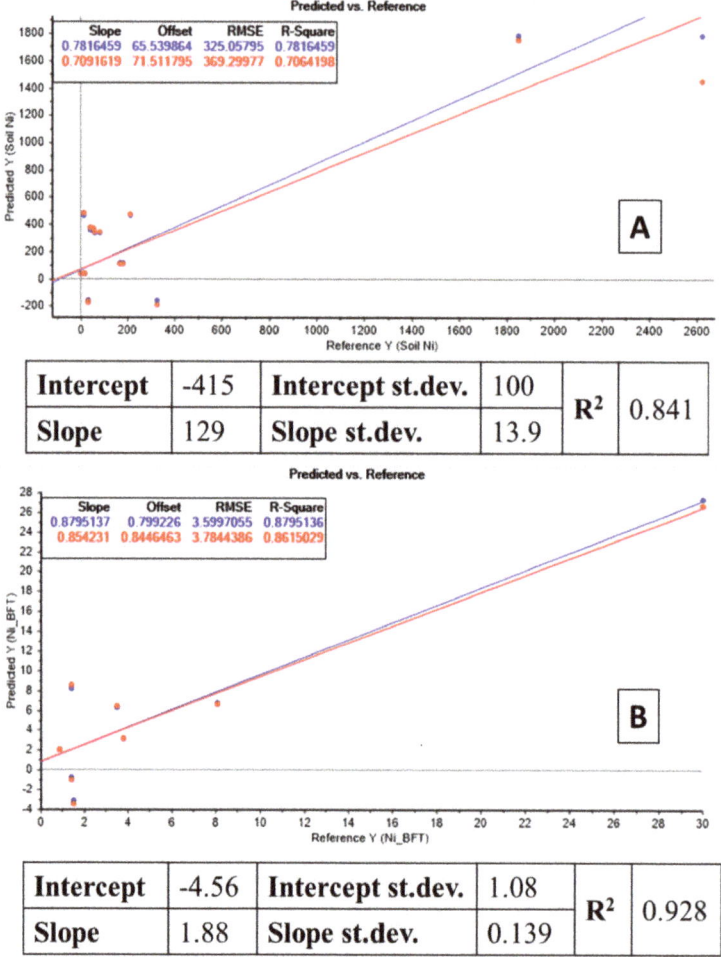

Figure 3. (**A**) Table: Linear regression between total Ni concentration in soil and Ni in *P. annua* shoots. Plot: recalculated total soil Ni by the model, input data derived from the linear relation between total Ni in soil, and Ni in plant. Blue dots: forecasted soil Ni concentrations in calibration mode (all soil data were used as input). Red dots: forecasted soil Ni concentrations in cross-validation mode, excluding one soil data at a time (leave-one-out mode). (**B**) Table 2: Linear regression between bioavailable Ni concentration in soil and Ni in *P. annua* shoots. Plot: Recalculated total soil Ni by the model; input data are derived from the linear relation between total Ni in soil and Ni in plant. Blue dots: forecasted soil Ni concentrations in calibration mode (all soil data were used as input). Red dots: forecasted soil Ni concentrations in cross-validation mode, excluding one soil data at a time (leave-one-out mode).

Table 3. Total and bioavailable concentrations of six heavy metals in the analyzed soils.

Soil	pH	OM(%)	Zn (ppm)		Cu (ppm)		Pb (ppm)		Cr (ppm)		Cd (ppm)		Ni (ppm)	
			Total	Bioavail.	Total	Bioavail.	Total	Bioavail.	Total	Bioavail.	Total	Bioavail.	Total	Bioavail.
MI3	7.73	13.1	1200 ± 300	N.A.	390 ± 30	36 ± 3	530 ± 50	28 ± 2	229 ± 10	0.48 ± 0.03	0.48 ± 0.01	0.03 ± 0.01	175 ± 5	3.80 ± 0.02
MI4	7.75	9.40	1200 ± 200	N.A.	540 ± 40	113 ± 7	135 ± 2	5.5 ± 0.3	75 ± 1	0.61 ± 0.06	0.39 ± 0.01	0.08 ± 0.01	70 ± 10	3.5 ± 0.1
MI9	8.81	9.48	270 ± 40	N.A.	60 ± 10	N.A.	22 ± 3	N.A.	100 ± 70	N.A.	0.40 ± 0.2	N.A.	60 ± 30	N.A.
BO7	9.04	7.90	410 ± 70	N.A.	133 ± 7	11 ± 1	110 ± 20	4.3 ± 0.4	40 ± 10	0.15 ± 0.01	0.29 ± 0.01	0.08 ± 0.01	39 ± 4	1.5 ± 0.1
BO8	8.91	6.59	510 ± 40	N.A.	110 ± 10	2.6 ± 0.2	120 ± 20	4.6 ± 0.4	150 ± 60	0.19 ± 0.01	0.6 ± 0.1	0.06 ± 0.01	130 ± 90	1.4 ± 0.1
NAT1	7.42	11.8	56 ± 3	N.A.	12 ± 2	2.6 ± 0.2	12 ± 4	2.8 ± 0.1	19 ± 1	0.83 ± 0.03	0.10 ± 0.05	0.19 ± 0.02	11 ± 3	0.9 ± 0.1
NAT5	8.79	4.57	184 ± 6	N.A.	50 ± 5	3.9 ± 0.4	19 ± 5	0.53 ± 0.01	22 ± 5	0.07 ± 0.01	0.33 ± 0.04	0.04 ± 0.01	49 ± 4	4.3 ± 0.3
NAT8	8.79	2.70	110 ± 10	N.A.	19 ± 2	N.A.	8 ± 4	N.A.	570 ± 50	N.A.	0.38 ± 0.03	N.A.	1900 ± 300	N.A.

From Table 3, it can be noted that Zn content is one order of magnitude higher in urban soils from Milan (>1000 ppm) if compared with other soils, while the lowest values can be found in woodland soils (about 100 ppm). The situation was similar for copper: urban soils from Milan had the highest values (>500 ppm) and woodland soils had lowest levels of Cu (~10–50 ppm). Again, for Pb, a similar situation to the previous two metals was detectable: the concentrations varied from approximately 100 to 500 ppm for Milan urban soils to approximately 10 to 20 ppm for woodland soils. These first three metals were therefore connected to anthropogenic activities. The situation of Cd is substantially different, which had similar levels (~0.35 ppm) in all the analyzed soils. Finally, Cr and Ni showed a wide range of concentrations in soils, independently of the origin of the soil. Nevertheless, the highest concentrations of these metals were found in the ultramafic soils from Prinzera (Ni: ~1800 ppm; Cr: ~500 ppm).

3.2. Species and Metal Selection

A preliminary exploration tested the correlation between the BAF calculated on TM (BAF_TM) and BAF calculated on BM (BAF_BM). This step was useful to evaluate if the response of our species was consistent both in considering the total metal in the soil and the bioavailable fraction. Only plants and metals that had high correlation values were kept as candidates for bioindication. Species with high correlation values for certain metals were likely to give consistent information on total and bioavailable metals in soil simultaneously.

High correlation values (Table 4) between BAF_TM and BAF_BM were found for all species in at least two metals each. From this preliminary screening, *S. vulgaris* appeared to be a possible candidate for bioindication of Cu, Pb, Cd, and Ni. *P. aviculare* was found to be a potential bioindicator of Pb, Cr, Cd, and Ni. Finally, *P. annua* could be a possible bioindicator for Pb and Ni.

Table 4. Correlation table between BAF_TM and BAF_BM. Colors indicate the "goodness" of correlation: yellow = significant; orange = relevant; green = high; blue = excellent. Information about Zn was not available.

Correlation BAF_TM/BAF_BM	Zn	Cu	Pb	Cr	Cd	Ni
P. annua	N.A.	0.33	0.79	−0.03	0.10	0.87
P. aviculare	N.A.	0.00	0.78	0.94	0.83	0.81
S. vulgaris	N.A.	0.97	1.00	−0.22	0.98	0.97

Metal concentrations for all plants are reported in Table 5.

Table 5. Metal concentrations and bioaccumulation factor (BAF) for the three studied species.

Species	Soil	Zn (ppm) Plant	BAF	Cu (ppm) Plant	BAF	Pb (ppm) Plant	BAF	Cr (ppm) Plant	BAF	Cd (ppm) Plant	BAF	Ni (ppm) Plant	BAF
S. vulgaris	Mi3	17.9 ± 0.7	0.01	0.78 ± 0.05	0.02	<LoD	N.D.	<LoD	N.D.	0.05 ± 0.01	0.02	0.39 ± 0.03	0.02
S. vulgaris	Mi4	471 ± 40	0.32	9.8 ± 0.9	0.39	<LoD	N.D.	0.16 ± 0.01	0.02	0.05 ± 0.01	0.02	0.67 ± 0.03	0.03
S. vulgaris	Mi9	70 ± 1	0.39	8.6 ± 0.5	0.28	0.36 ± 0.02	0.09	0.51 ± 0.01	0.16	0.21 ± 0.02	0.16	0.64 ± 0.03	0.02
S. vulgaris	Bo7	99 ± 4	0.17	7.3 ± 0.5	0.07	0.77 ± 0.02	0.01	0.49 ± 0.02	0.02	0.21 ± 0.01	2.49	1.25 ± 0.01	0.04
S. vulgaris	Bo8	5.4 ± 0.5	0.01	0.08 ± 0.01	0.01	<LoD	N.D.	0.02 ± 0.01	0.02	0.06 ± 0.01	0.09	0.25 ± 0.02	0.02
S. vulgaris	Nat8	17 ± 1	0.16	2.9 ± 0.2	0.16	0.48 ± 0.03	0.03	1.14 ± 0.07	0.03	0.73 ± 0.02	1.86	5.1 ± 0.3	0.02
S. vulgaris	Nat1	54 ± 4	1.02	8.26 ± 0.03	0.69	0.42 ± 0.03	0.02	0.7 ± 0.01	0.05	0.76 ± 0.03	0.01	3.07 ± 0.06	0.06
S. vulgaris	Nat5	22.7 ± 0.6	0.13	5.7 ± 0.2	0.14	4.96 ± 0.03	0.31	0.16 ± 0.01	0.01	14.6 ± 0.4	0.02	2.4 ± 0.1	0.06
P. aviculare	Mi3	47 ± 3	0.03	14 ± 1	0.04	0.82 ± 0.03	0.02	1.7 ± 0.1	0.01	0.17 ± 0.01	0.36	1.93 ± 0.08	0.01
P. aviculare	Mi4	56 ± 2	0.04	13.7 ± 0.7	0.03	1.03 ± 0.08	0.01	1.58 ± 0.08	0.02	0.17 ± 0.01	0.44	0.62 ± 0.01	0.01
P. aviculare	Mi9	30 ± 3	0.16	21 ± 1	0.22	0.63 ± 0.05	0.02	3.52 ± 0.07	0.13	0.24 ± 0.01	1.86	1.8 ± 0.2	0.15
P. aviculare	Bo7	57 ± 3	0.13	19.8 ± 0.2	0.15	0.81 ± 0.01	0.01	2.5 ± 0.2	0.12	0.47 ± 0.04	5.48	0.65 ± 0.03	0.02
P. aviculare	Bo8	46 ± 5	0.20	7.8 ± 0.7	0.14	1.02 ± 0.08	0.01	1.3 ± 0.1	0.08	4.04 ± 0.07	1.72	1.23 ± 0.05	0.02
P. aviculare	Nat8	32 ± 2	0.31	39 ± 1	2.17	0.12 ± 0.01	0.01	0.22 ± 0.03	0.02	0.16 ± 0.01	0.42	2.2 ± 0.2	0.02
P. aviculare	Nat1	40 ± 2	0.75	3.6 ± 0.1	0.30	0.11 ± 0.02	N.D.	0.17 ± 0.01	0.03	0.35 ± 0.01	5.01	1.2 ± 0.1	0.06
P. aviculare	Nat5	27.4 ± 0.6	0.29	3.15 ± 0.09	0.14	0.13 ± 0.01	N.D.	0.12 ± 0.01	0.01	0.33 ± 0.02	1.43	1.6 ± 0.2	0.03
P. annua	Mi3	220 ± 10	0.15	1.80 ± 0.01	0.01	<LoD	N.D.	<LoD	N.D.	0.35 ± 0.01	0.71	4.1 ± 0.3	0.03
P. annua	Mi4	108 ± 8	0.20	14.0 ± 0.6	0.07	0.08 ± 0.01	0.01	2.33 ± 0.01	0.07	0.46 ± 0.04	1.92	5.8 ± 0.4	0.07
P. annua	Mi9	84.0 ± 0.4	0.47	14.2 ± 0.1	0.15	0.54 ± 0.02	0.02	4.14 ± 0.05	0.14	0.42 ± 0.02	3.31	6.8 ± 0.6	0.20
P. annua	Bo7	29 ± 2	0.10	11.0 ± 0.8	0.08	0.17 ± 0.03	0.01	0.45 ± 0.02	0.01	0.13 ± 0.01	0.48	0.82 ± 0.01	0.06
P. annua	Bo8	119 ± 9	0.26	20.2 ± 0.3	0.19	0.25 ± 0.02	0.01	1.52 ± 0.06	0.07	0.23 ± 0.02	0.48	2.00 ± 0.07	0.14
P. annua	Nat8	4.0 ± 0.4	0.04	4.1 ± 0.4	0.23	0.54 ± 0.02	0.03	0.38 ± 0.01	0.01	1.53 ± 0.01	3.89	17.2 ± 0.2	0.02
P. annua	Nat1	98 ± 5	1.85	5.2 ± 0.3	0.43	1.7 ± 0.1	0.08	3.1 ± 0.1	0.21	0.61 ± 0.06	8.57	3.5 ± 0.3	0.19
P. annua	Nat5	34 ± 1	0.37	7.7 ± 0.7	0.33	0.41 ± 0.03	0.01	1.6 ± 0.1	0.06	0.23 ± 0.01	1.01	6.27 ± 0.03	0.17

These data were used to carry out another explorative analysis focused on the correlation between plant metal content (PM) and the soil metal content (TM). High correlations were found for two species, but only in the case of Ni (Table 6); therefore, the potential of *P. annua* and *S. vulgaris* as bioindicators for Ni was further explored and modeled.

Table 6. Correlation table between PM and TM. Colours indicate the "goodness" of correlation: yellow = significant; orange = relevant; green = high; blue = excellent.

Correlation TM/PM	Zn	Cu	Pb	Cr	Cd	Ni
P. annua	0.64	0.15	−0.47	−0.47	−0.05	0.87
P. aviculare	0.59	−0.05	0.50	−0.27	0.61	0.62
S. vulgaris	0.56	0.04	−0.28	0.54	−0.02	0.73

Some other relevant correlations were found in *P. aviculare* for Cd and Pb and for *P. annua* for Cr, but these species did not show any linear relation with the metal when furtherly tested. Hence, attention is focused on Ni in the following section, with the aim of creating predictive models for Ni bioindication using *P. annua* and *S. vulgaris*.

3.3. Bioindication of Ni Using P. annua

The strong linear relation ($R^2 = 0.841$) between total Ni in soil and Ni concentration in *P. annua* shoots is shown in the table enclosed in Figure 3A. This linear relation made it possible the creation of a MLR model with predictive potential (response plot in Figure 3A). The performance of the models was considered relevant to bioindication purposes since this model appeared reliable when tested by ANOVA (p-value related to the F parameter <0.05). Both in calibration (blue dots) and cross-validation (red dots), the two lines and the values almost overlapped. Except for high Ni values, the model appeared very accurate in the prevision of total Ni in soil using as input data Ni concentration measured in *P. annua* plants.

An even better relation ($R^2 = 0.928$) was obtained when considering the bioavailable fraction of Ni compared to Ni content in *P. annua* shoots (table enclosed in Figure 3B). The connected MLR model showed an even higher predictive potential, with high accuracy for the whole range of Ni values (response plot in Figure 3B). The high R values and the solidity of both models confirmed that *P. annua* can be used as reliable Ni bioindicator.

3.4. Bioindication of Ni Using S. vulgaris

The situation of *S. vulgaris* is similar to the one of *P. annua*. The data showed a clear linear relation ($R^2 = 0.908$) between Ni in soil and Ni in plant (table enclosed in Figure 4A). From this strong relation, the creation of a predictive MLR model was also possible (response plot in Figure 4A). For low Ni values the performance of the models was high both in calibration and cross-validation mode. While for high Ni values, results obtained in cross-validation were slightly different from the calibrations ones. The model appeared reliable when tested by ANOVA (p-value related to the F parameter <0.05) therefore significant for bioindication purposes.

An even better linear relation ($R^2 = 0.969$) was obtained when considering the bioavailable Ni pool in soil compared to *P. annua* Ni content (table enclosed in Figure 4B). The connected MLR model showed therefore an even higher predictive potential (response plot in Figure 4B). In this case, for all Ni concentrations the performance of the model was similar, with the data forecasted in cross-validation almost overlapped to the ones calculated in calibration. This high similarity of the two sets gave strength to the model, which can be considered highly reliable for available Ni prevision in soil.

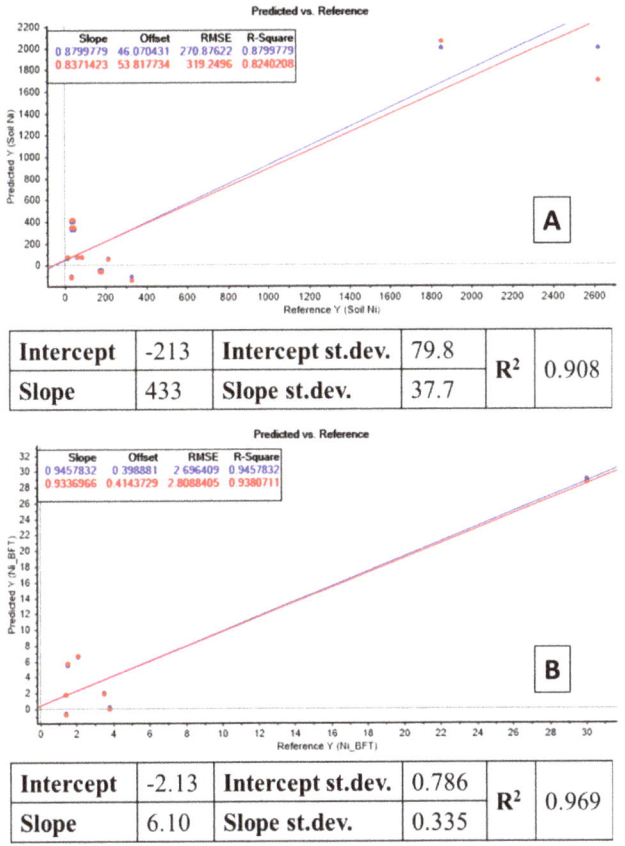

Figure 4. (**A**) Table: Linear regression between total Ni concentration in soil and Ni in *S. vulgaris* shoots. Plot: recalculated total soil Ni by the model, input data derived from the linear relation between total Ni in soil and Ni in plant. Blue dots: forecasted soil Ni concentrations in calibration mode (all soil data were used as input). Red dots: forecasted soil Ni concentrations in cross-validation mode, excluding one soil data at a time (leave-one-out mode). (**B**) Table: Linear regression between bioavailable Ni concentration in soil and Ni in *S. vulgaris* shoots. Plot: Recalculated total soil Ni by the model; input data are derived from the linear relation between total Ni in soil and Ni in plant. Blue dots: forecasted soil Ni concentrations in calibration mode (all soil data were used as input). Red dots: forecasted soil Ni concentrations in cross-validation mode, excluding one soil data at a time (leave-one-out mode).

4. Discussion

The analyzed soils in this work were selected from three different environments characterized by different levels and sources of HMs pollution. Urban soils from Milan and Bologna, despite some peculiarities connected to parent materials that contributed to their pedogenesis, were characterized by similar amounts and types of HMs. All urban surfaces, in fact, receive deposits that mainly come from anthropic activities, like vehicle emissions, industrial discharges, domestic heating, and material weathering [41,42].

Street dust and top roadside soils in urban areas are typical sinks of HMs from atmospheric deposition and water runoff. Key HMs in these zones are Pb from gasoline additives; Cu, Zn, and Cd from car components; tire abrasion; lubricants; and industrial emissions [43,44]. MI3 and MI4

(Table 4) had the most polluted soils as they were collected at a very busy street crossing. This relation underlined the important role of vehicular traffic in contributing to soil pollution.

Woodland soils were collected in natural areas that were not influenced by anthropic activities; trace elements detected in such soils were the one originally present in parent materials. However, a small contribution form diffuse sources of HMs pollution, like vehicles emission, cannot be excluded for these samples. It has been reported that fine particulate for example coming from tires and brakes abrasion may fall far away from source location [45]. Finally, ultramafic soil, despite being collected far from anthropic source of pollution, is still present at physiologic high levels of Ni and Cr. The level of anthropic HMs (Zn, Cu, Pb, Cd) was low, demonstrating the naturalness of this environment, but the ultramafic rock on which pedogenesis took place was highly enriched in Ni and Cr, now found in soils [46].

The marked heterogeneity of sampled soils was the main obstacle to the creation plant-soil linear relations in the absorption of metals. The investigated soils showed a wide range of properties (like different texture, pH, OM content, etc.) that deeply affected the bioavailability of metal to plants [47,48].

Moreover, it is widely known that trace elements are not found in plant tissues with the same the proportion of their concentrations in soil [49]. The uptake of these elements by plants is selective: essential nutrients (like Zn, Cu, and Mn) are actively taken and show a more linear relation to soil concentration if compared to nonessential nutrient [47,50]. This active absorption of micronutrients eventually results in a greater translocations and concentration in plant shoots, but also to a higher toxicity if compared to nonessential nutrient [51].

In agreement with these uptake mechanisms, the species investigated in this paper showed higher correlations for Zn and Ni (micronutrients) than for nonessential ones (Table 6). Despite the correlation coefficient for Zn of about 0.6 for all species (Table. 6), no linear relations were found for this metal. On the other hand, *P. annua* ($R^2 = 0.87$) and *S. vulgaris* ($R^2 = 0.73$) showed a linear relation with Ni in soil, as similarly found for *Taraxacum officinale* [25].

Another possible reason for the lack of linear relations between metal content in soils and plants was probably due to the poor translocation of these elements from root to shoot. Other studies in fact demonstrated that when nonessential metals are present at high concentration in soil, most herbaceous plants tend to use exclusion strategies to prevent the uptake of these toxic elements [52,53]. This phenomenon was observed for Cd in *Halophyla ovalis* [51] and for Pb [54]. Species unable to prevent root absorption, instead limit the translocation to shoots keeping the majority of toxic elements stored in roots [55].

In the present study roots were not collected, so data about metal concentration in belowground organs were not available, but an extensive literature demonstrated that root metal concentration better correlates with soil HM concentration if compared to shoots [56,57]. However, shoots are the most used parts for bioindication purposes, due to their visibility and easiness of collection. For this reason, the use of plants with limited translocation of HMs in aboveground parts has low practical application [56].

The bioaccumulation factor (BAF), a parameter that quantifies the element transfer from soil to plant, was found to be lower than one for almost all samples in reference to all metals. Due to this low BAF, all the studied species can be considered nonaccumulators, according to van der Ent et al. [58] metal accumulator plants must have this parameter always higher than one. Interestingly, several BAF values were found to be above one for Cd (Table 5), and similar results were also reported for *P. major* by Galal & Shehata [23], demonstrating that this metal at low concentrations can be easily uptaken and transferred to aerial parts.

P. annua and *S. vulgaris* (Figure 5) had similar BAF values for Ni uptake in all soils, making them suitable for bioindication purposes.

Figure 5. (**A**) *Senecio vulgaris* growing at a busy street crossing in Bologna. (**B**) *Poa annua* growing on the sidewalk in Milan.

This is in line with the guidelines from the EPA [59], in which it is stated that good indicator plants should keep this parameter constant in several soil conditions. Moreover, Ni is very mobile inside the plant, and is transported (binded by organic acids) through the xylematic flow from roots to shoots [60]. This work demonstrated that aerial parts of the common weeds *P. annua* and *S. vulgaris* can be used as environmental indicators of Ni pollution in soil. Similar results were achieved for *Urtica dioica, Taraxacum officinale, Plantago major*, and two Trifolium species for Pb, Mn, and Cu [20]. The use of common weeds can be a valid alternative to the use of lichens in assessing HMs levels in cities, especially because these herbaceous plants are common and easily recognizable.

Interestingly, Malizia et al. [20] achieved similar results when assessing HMs in soil using lichens or using herbaceous plants for Cu, Zn and Pb, in the city of Rome. This promising result should encourage the research on common weeds as valid alternative to "lichens biomonitoring" in urban areas.

Despite the extensive literature about biomonitoring using herbaceous species [20,61], most studies focused on *Taraxacum officinale* while only few took into account other species [20].

This study results highlighted the possibility to find new species suitable for bioindication of metal pollution in anthropic environment. The importance of having several bioindicator species in each environment has been underlined by Phillips et al. [57] who suggested a "multi-species" approach to bioindication, in order to obtain more precise results. Finally, we demonstrated the possibility to create predictive models when strong linear relations are present between soil Ni and plant Ni. This chemometric approach was not aimed at the replacement of collection and analysis of samples, but, instead, to give support at the results achieved with traditional field samplings and lab analysis.

5. Conclusions

Our results about the possible use of *S. vulgaris, P. aviculare*, and *P. annua*, as HMs bioindicators, showed that metal concentrations in soils and plants mostly do not correlate under natural growth conditions. Despite metals found in soils being present in plants, the concentration in aboveground organs is deeply influenced by soil properties and plant translocation. For Zn, Cu, Pb, Cr, and Cd, none of the studied species was enough equipped to be a good bioindicator; moreover, *P. aviculare* was found to be inappropriate even as Ni indicator. Nevertheless, the present work demonstrated the feasibility to use *P. annua* and *S. vulgaris* as bioindicators of Ni pollution in soil. The two species were reliable indicators of total and bioavailable Ni fraction. The good results achieved with this two species allowed us to develop models based on our data, which were able to forecast with great accuracy Ni concentration in soil from Ni in plants. However, these promising results need a definitive validation using a greater number of samples.

Author Contributions: M.S. collected samples, analyzed data, and contributed to writing the manuscript and to the coordination of the study; D.M. and A.Z. analyzed the data and contributed to writing the manuscript; S.C. and F.d.L performed the analyses by AAS; A.T. coordinated the study and contributed to writing the manuscript. All authors revised the manuscript.

Funding: This research received no external funding

Conflicts of Interest: The authors declare no conflicts of interest.

Abbreviations:

Atomic Absorption Spectroscopy = AAS;
Bioaccumulation Factor = BAF;
Bioaccumulation Factor calculated on bioavailable metal in soil = BAF_BM;
Bioaccumulation Factor calculated on total metal in soil = BAF_TM;
Bioavailable soil metal concentration = BM;
Heavy metals = HMs;
Inorganic matter = IM;
leave-one-out cross-validation = LOO-CV;
Limit of detection = LoD;
Organic matter = OM; Plant metal concentration = PM;
Total soil metal concentration = TM.

References

1. Wang, Z.; Yao, L.; Liu, G.; Liu, W. Heavy metals in water, sediments and submerged macrophytes in ponds around the Dianchi Lake. China. *Ecotox. Environ. Saf.* **2014**, *107*, 200–206. [CrossRef] [PubMed]
2. Ahmad, S.S.; Reshi, Z.A.; Shah, M.A.; Rashid, I.; Ara, R.; Andrabi, S.M.A. Heavy metal accumulation in the leaves of Potamogeton natans and Ceratophyllum demersum in a Himalayan RAMSAR site: Management implications. *Wetlands Ecol. Manag.* **2016**, *24*, 469–475. [CrossRef]
3. Charlesworth, S.; De Miguel, E.; Ordóñez, A. A review of the distribution of particulate trace elements in urban terrestrial environments and its application to considerations of risk. *Environ. Geochem. Health* **2011**, *33*, 103–123. [CrossRef] [PubMed]
4. Melucci, D.; Locatelli, M.; Locatelli, C.; Zappi, A.; De Laurentiis, F.; Carradori, S.; Campestre, C.; Leporini, L.; Zengin, G.; Picot, C.M.N.; et al. A comparative assessment of biological effects and chemical profile of Italian asphodeline lutea extracts. *Molecules* **2018**, *23*, 461. [CrossRef] [PubMed]
5. Tchounwou, P.B.; Yedjou, C.G.; Patlolla, A.K.; Sutton, D.J. Heavy metal toxicity and the environment. In *Molecular, Clinical and Environmental Toxicology*; Luch, A., Ed.; Springer: New York, NY, USA, 2012; p. 573.
6. He, Z.L.; Yang, X.E.; Stoffella, P.J. Trace elements in agroecosystems and impacts on the environment. *J. Trace Elem. Med. Biol.* **2005**, *19*, 125–140. [CrossRef] [PubMed]
7. Li, J.; Lu, Y.; Yin, W.; Gan, H.; Zhang, C.; Deng, X.; Deng, X.; Lian, J. Distribution of heavy metals in agricultural soils near a petrochemical complex in Guangz-hou, China. *Environ. Monit. Assess.* **2009**, *153*, 365–375. [CrossRef]
8. Szyczewski, P.; Siepak, J.; Niedzielski, P.; Sobczyński, T. Research on heavy metals in Poland. *Pol. J. Environ. Stud.* **2009**, *5*, 755–768.
9. Kabata-Pendias, A.; Mukherjee, A.B. *Trace Elements from Soil to Human*; Springer: Berlin, Germany, 2007.
10. Norgate, T.E.; Jahanshahi, S.; Rankin, W.J. Assessing the environmental impact of metal production processes. *J. Clean. Prod.* **2007**, *15*, 838–848. [CrossRef]
11. Markert, B.; Wunschmann, S.; Franzle, S.; Graciana Figuereido, A.M.; Ribeirao, A.; Wang, M. Bioindication of atmospheric trace metals-with special reference to megacities. *Environ. Pollut.* **2011**, *159*, 1991–1995. [CrossRef]
12. Wiseman, C.L.S.; Zereini, F.; Puttmann, W. Traffic-related trace element fate and uptake by plants cultivated in roadside soils in Toronto, Canada. *Sci. Total Environ.* **2013**, *442*, 86–95. [CrossRef]
13. Zereini, F.; Wiseman, C.L.S.; Puttmann, W. Changes in palladium, platinum, and rhodium concentrations, and their spatial distribution in soils along a major highway in Germany from 1994 to 2004. *Environ. Sci. Technol.* **2007**, *41*, 451–456. [CrossRef] [PubMed]

14. Lough, G.; Schauer, J.J.; Park, J.S.; Shafer, M.M.; Deminter, J.T.; Weinstein, J. Emissions of metals associated with motor vehicle roadways. *Environ. Sci. Technol.* **2005**, *39*, 826–836. [CrossRef] [PubMed]
15. Zereini, F.; Alsenz, H.; Wiseman, C.L.S.; Puttmann, W.; Reimer, E.; Schleyer, R.; Bieber, E.; Wallasch, M. Platinum group elements (Pt, Pd, Rh) in airborne particulate matter in rural vs. urban areas of Germany: Concentrations and spatial patterns of distribution. *Sci. Total Environ.* **2012**, *416*, 261–268. [CrossRef] [PubMed]
16. Onder, S.; Dursun, S. Air borne heavy metal pollution of Cedrus libani (A. Rich) in the city centre of Konya (Turkey). *Atmos. Environ.* **2006**, *40*, 1122–1133. [CrossRef]
17. Jozic, M.; Peer, T.; Turk, R. The impact of the tunnel exhausts in terms of heavy metals to the surrounding ecosystem. *Environ. Monit. Assess.* **2009**, *150*, 261–271. [CrossRef]
18. Locatelli, C.; Melucci, D. Voltammetric determination of ultra-trace total mercury and toxic metals in meals. *Food. Chem.* **2012**, *30*, 460–466. [CrossRef]
19. Locatelli, C.; Melucci, D. Voltammetric method for ultra-trace determination of total mercury and toxic metals in vegetables. Comparison with spectroscopy. *Cent. Eur. J. Chem.* **2013**, *11*, 790–800. [CrossRef]
20. Malizia, D.; Giuliano, A.; Ortaggi, G.; Masotti, A. Common plants as alternative analytical tools to monitor heavy metals in soil. *Chem. Cent. J.* **2012**, *6*, S6. [CrossRef]
21. Jenkis, D.A. Trace elements in saxicolous lichens. In *Pollutant Transport and Fate in Ecosystems*; Coughtrey, P.J., Martin, M.H., Unsworth, M.H., Eds.; Blackwell Sci.: Oxford, UK, 1987; p. 422.
22. Jiang, Y.; Fan, M.; Hu, R.; Zhao, J.; Wu, J. Mosses are better than leaves of vascular plants in monitoring atmospheric heavy metal pollution in urban areas. *Int. J. Environ. Res. Public Health* **2018**, *15*, 1105. [CrossRef]
23. Galal, T.M.; Shehata, H.S. Bioaccumulation and translocation of heavy metals by Plantago major L. grown in contaminated soils under the effect of traffic pollution. *Ecol. Indic.* **2015**, *48*, 244–251. [CrossRef]
24. Rule, J.H. Use of small plants as trace elements phytomonitors, with emphasis on the common dandelion, taraxacum officinale. In *Biogeochemistry of Trace Element*; Adriano, D.C., Chen, Z.S., Yang, S.S., Iskandar, I.K., Eds.; Science Reviews: Wales, UK, 1997; 432p.
25. Fröhlichová, A.; Száková, J.; Najmanová, J.; Tlustoš, P. An assessment of the risk of element contamination of urban and industrial areas using Taraxacum sect. Ruderalia as a bioindicator. *Environ. Monit. Assess.* **2018**, *190*, 150–163. [CrossRef]
26. Wittig, R. General aspects of biomonitoring heavy metals by plants. In *Plants as Biomonitors*; Market, B., Ed.; VCH: Weinheim, Germany, 1993.
27. Lin, Y.X.; Zhang, X.M. Accumulation of heavy metals and the variation of amino acids and protein in Eichhornia crassipes (Mart.) Solms in the Dianchi Lake. *Oceanol. Limnol. Sinica* **1990**, *21*, 179–184.
28. Keane, B.; Collier, M.; Shann, J.R.; Rogstad, S.H. Metal content of dandelion (Taraxacum officinale) leaves in relation to soil contamination and airborne particulate matter. *Sci. Total Environ.* **2001**, *281*, 63–78. [CrossRef]
29. Karlsson, N. On molybdenum in Swedish soil and vegetation and some related questions. *Medd. 23 Stat. LantbrKem. Kontrollanst. Stockholm* **1961**, 234–247.
30. Kashem, M.A.; Singh, B.R. Metal availability in contaminated soils: I. Effects of flooding and organic matter on changes in Eh, pH and solubility of Cd, Ni and Zn. *Nutr. Cycl. Agroecosyst.* **2001**, *61*, 247–255. [CrossRef]
31. Antoniadis, V.; Robinson, J.S.; Alloway, B.J. Effects of short-term pH fluctuations on cadmium, nickel, lead, and zinc availability to ryegrass in a sewage sludge-amended field. *Chemosphere* **2008**, *71*, 759–764. [CrossRef]
32. Dai, J.; Becquer, T.; Rouiller, J.H.; Reversat, G.; Reversat, F.B.; Lavelle, P. Influence of heavy metals on C and N mineralization and microbial biomass in Zn-, Pb-, Cu-, and Cd-contaminated soils. *Appl. Soil Ecol.* **2004**, *25*, 99–109. [CrossRef]
33. Ghosh, M.; Singh, S.P. A review on phytoremediation of heavy metals and utilization of its byproducts. *Appl. Ecol. Environ. Res.* **2005**, *3*, 1–18. [CrossRef]
34. Nanda, S.; Abraham, A. Remediation of heavy metal contaminated soil. *Afr. J. Biotechnol.* **2013**, *12*, 3099–3109.
35. Huang, C.Y.; Schulte, E.E. Digestion of plant tissue for analysis by ICP emission spectroscopy. *Commun. Soil Sci. Plant Anal.* **1985**, *16*, 943–958. [CrossRef]
36. Quevauviller, P. Operationally-defined extraction procedures for soil and sediment analysis. Part 3: New CRMs for trace-element extractable contents. *TrAC* **2002**, *21*, 774–785. [CrossRef]
37. Miller, J.N.; Miller, J.C. *Statistics and Chemometrics for Analytical Chemistry*; Pearson Education: London, UK, 2010.

38. Walkley, A.; Black, I.A. An examination of the Degtjareff method for determining soil organic matter, and a proposed modification of the chromic acid titration method. *Soil Sci.* **1934**, *37*, 29–38. [CrossRef]
39. Storer, D.A. A simple high sample volume ashing procedure for determining soil organic matter. *Commun. Soil Sci. Plant Anal.* **1984**, *15*, 759–772. [CrossRef]
40. Brereton, R.G. *Applied Chemometrics for Scientists*; Wiley: Hoboken, NJ, USA, 2007.
41. Gibson, M.G.; Farmer, J.G. Multi-step chemical extraction of heavy metals from urban soils. *Environ. Pollut. B* **1986**, *11*, 117–135. [CrossRef]
42. Kelly, J.; Thornton, I.; Simpson, P.R. Urban geochemistry: A study of influence of anthropogenic activity on heavy metal content of soils in traditionally industrial and non-industrial areas of Britain. *Appl. Geochem.* **1996**, *11*, 363–370. [CrossRef]
43. Markus, J.A.; McBratney, A.B. An urban soil study: Heavy metals in Glebe, Australia. *Aust. J. Soil Res.* **1996**, *34*, 453–465. [CrossRef]
44. Wilcke, W.; Muller, S.; Kanchanakool, N.; Zech, W. Urban soil contamination in Bangkok: Heavy metal and aluminium portioning in topsoils. *Geoderma* **1998**, *86*, 211–228. [CrossRef]
45. Hulskotte, J.H.; van der Gon, H.A.; Visschedijk, A.J.; Schaap, M. Brake wear from vehicles as an important source of diffuse copper pollution. *Water Sci. Technol.* **2007**, *56*, 223–231. [CrossRef]
46. Kierczak, J.; Neel, C.; Bril, H.; Puziewicz, J. Effect of mineralogy and pedoclimatic variations on Ni and Cr distribution in serpentine soils under temperate climate. *Geoderma* **2007**, *142*, 165–177. [CrossRef]
47. Kabata-Pendias, A. *Trace Elements in Soils and Plants*, 4th ed.; Taylor & Francis Group: Milton Park, UK, 2011.
48. Yang, J.; Ye, Z. Metal accumulation and tolerance in wetland plants. *Front. Biol. China* **2009**, *4*, 282–288. [CrossRef]
49. Curlık, J.; Kolesar, M.; Durza, O.; Hiller, E. Dandelion (Taraxacum officinale) and Agrimony (Agrimonia eupatoria) as Indicators of Geogenic Contamination of Flysch Soils in Eastern Slovakia. *Arch. Environ. Contam. Toxicol.* **2016**, *70*, 475–486. [CrossRef]
50. Smillie, C. *Salicornia* spp. as a biomonitor of Cu and Zn in salt marsh sediments. *Ecol. Indic.* **2015**, *56*, 70–78. [CrossRef]
51. Ralph, P.J.; Burchett, M.D. Photosynthetic responses of Halophila ovalis to heavy metals stress. *Environ. Pollut.* **1998**, *103*, 91–101. [CrossRef]
52. Weis, J.S.; Glover, T.; Weis, P. Interactions of metals affect their distribution in tissues of Phragmites australis. *Environ. Poll.* **2004**, *131*, 409–415. [CrossRef]
53. Weis, J.S.; Weis, P. Metal uptake, transport and release by wetland plants: Implications for phytoremediation and restoration. *Environ. Int.* **2004**, *30*, 685–700. [CrossRef]
54. Sharma, P.; Dubey, R.S. Lead toxicity in plants. *Braz. J. Plant Physiol.* **2005**, *17*, 35–52.
55. Stoltz, E.; Greger, M. Accumulation properties of As, Cd, Cu, Pb and Zn by four wetland plant species growing on submerged mine tailings. *Environ. Exp. Bot.* **2002**, *47*, 271–280. [CrossRef]
56. Llagostera, I.; Pérez, M.; Romero, J. Trace metal content in the seagrass Cymodocea nodosa: Differential accumulation in plant organs. *Aquat. Bot.* **2011**, *95*, 124–128. [CrossRef]
57. Phillips, D.P.; Human, L.R.D.; Adams, J.B. Wetland plants as indicators of heavy metal contamination. *Marine Pollut. Bull.* **2015**, *92*, 227–232. [CrossRef]
58. Van der Ent, A.; Baker, A.J.M.; Reeves, R.D.; Pollard, A.J.; Schat, H. Hyperaccumulators of metal and metalloid trace elements: Facts and fiction. *Plant Soil* **2012**, *362*, 319–334. [CrossRef]
59. EPA. *Framework for Metal Risk Assessment*; U.S Environmental Protection Agency; Office of the Science Advisor: Washington, DC, USA, 2007.
60. Yusuf, M.; Fariduddin, Q.; Hayat, S.; Ahmad, A. Nickel: An overview of uptake, essentiality and toxicity in plants. *Bull. Environ. Contam. Toxicol.* **2011**, *86*, 1–17. [CrossRef]
61. Kleckerová, A.; Dočekalová, H. Dandelion Plants as a Biomonitor of Urban Area Contamination by Heavy Metals. *Int. J. Environ. Res.* **2014**, *8*, 157–164.

Sample Availability: All samples are available from the authors.

 © 2019 by the authors. Licensee MDPI, Basel, Switzerland. This article is an open access article distributed under the terms and conditions of the Creative Commons Attribution (CC BY) license (http://creativecommons.org/licenses/by/4.0/).

Article

Assessment of the Quality of Polluted Areas in Northwest Romania Based on the Content of Elements in Different Organs of Grapevine (*Vitis vinifera* L.)

Florin Dumitru Bora [1], Claudiu Ioan Bunea [2], Romeo Chira [3] and Andrea Bunea [4,*]

[1] Research Station for Viticulture and Enology Târgu Bujor, Department of Physico-Chemistry and Biochemistry, 805200 Târgu Bujor, Romania; boraflorindumitru@gmail.com
[2] University of Agricultural Sciences and Veterinary Medicine, Department of Horticulture and Landscaping, 3-5 Mănăștur Street, 400372 Cluj-Napoca, Romania; claus_bunea@yahoo.com
[3] University of Medicine and Pharmacy "Iuliu Hatieganu", 3–5 Clinicilor Street, 400006 Cluj-Napoca, Romania; romeochira@yahoo.com
[4] University of Agricultural Sciences and Veterinary Medicine, Department of Chemistry, 3-5 Mănăștur Street, 400372 Cluj-Napoca, Romania
* Correspondence: andrea.bunea@usamvcluj.ro; Tel.: +40-264-596-384 (ext. 126)

Received: 19 January 2020; Accepted: 6 February 2020; Published: 9 February 2020

Abstract: The purpose of this study was to evaluate the environmental quality of polluted areas near the Baia Mare Mining and Smelting Complex for future improvements the quality of the environment in polluted areas, such as the city of Baia Mare and its surroundings. Samples of soil and organs of grapevine (*Vitis vinifera* L.) were collected from Baia Mare, Baia Sprie and surrounding areas (Simleul Silvaniei) and their content of Cu, Zn, Pb, Cd, Ni, Co, As, Cr, Hg were analyzed. Most soil and plant samples showed higher metal concentrations in Baia Mare and Baia Sprie areas compared to Simleul Silvaniei, exceeding the normal values. The results obtained from the translocation factors, mobility ratio, as well as from Pearson correlation study confirmed that very useful information is recorded in plant organs: root, canes, leaves and fruit. Results also indicated that *Vitis vinifera* L. has some highly effective strategies to tolerate heavy metal-induced stress, may also be useful as a vegetation protection barrier from considerable atmospheric pollution. At the same time, berries are safe for consumption to a large degree, which is a great advantage of this species.

Keywords: heavy metals; grapevine; bioaccumulation; biomonitoring

---

1. Introduction

Pollution is a worldwide problem caused by anthropogenic activities such as mining, petrochemical refining, and smelting, with negative impacts on human health. In Romania, 18% of population was exposed to heavy environmental pollution whereby serious health risks are likely. A total of 14 environmental pollution "hot spots" have been identified in Romania: Copșa Mică, Baia Mare, Ploiești-Brazi, Zlatna, Onești, Bacău, Suceava, Petești, Târgu Mures, Turnu Măgurele, Talcea, Isalnita, Brașov, and Govora; 5.3% of the population lives in these heavily polluted areas, mostly in the critical rural/urban interfaces [1–3].

Growing in extremely polluted areas, some plant species can be seriously damaged, whereas others can survive without any visible changes [4]. Uptake of trace metals by plants can happen from the soil through the roots and subsequent transport to the leaves or directly from the air. Specific mechanisms allow plant tissues to accumulate high quantities of trace metals, playing, thereby, a vital role in the natural recovery of industrial damage [5]. In this respect, trees are especially useful because contaminants can accumulate in their large biomass and they can grow in soil with poor fertility and structure [6]. Terrestrial higher plants are specific living-system structures with unique ecobiological characteristics. They interact actively with three spheres: soil, water and air, at the same time, requiring only modest nutrient input. Along with nutrients, plant roots can absorb a range of anthropogenic toxic materials. Heavy metals are just a class of such pollutants and several of them are well known as nonessential and extremely toxic for plants: cadmium (Cd), lead (Pb), mercury (Hg) and arsenic (As). Even essential micronutrients such as copper (Cu), zinc (Zn), and nickel (Ni) may become toxic for plants when absorbed above certain threshold values [7].

Plants have developed effective detoxification mechanisms to manage heavy metal content [8]. Some species may concentrate heavy metals in root cell walls and/or vacuoles, thus minimizing their phytotoxicity [9] and also preventing the spread of these contaminants in soil [10]. Phytoremediation is an excellent opportunity for cleaning up the pollute environment in an economic and ecological friendly manner. It uses green plants to detoxify the polluted environment, and it may be applied in a variety of ways [10]. On the other hand, plants can be used as indicators of the pollution level of the environment. Heavy metals in plant organs, especially in roots and leaves, represent a very specific evidence of spatial and temporal history of polluted area [11]. Researchers agree that the root and leaf analyses are essential in the evaluation process of the environmental quality of ecosystems or to study the effects of heavy metals on the chemical composition of plants.

Grapevine is an important crop worldwide, while the wine sector is of major importance for the economy of many countries [12]. The soil chemistry in vineyards influences wine and grape quality, vine-soil relationship being a key part of the concept of terroir [13,14]. The town of Baia Mare has been an important nonferrous metallurgical center where heavy metals like Pb and Cu have been extracted and processed for centuries. Metallurgical plants "Romplumb", located in the Ferneziu district, and "Cuprom", located in the eastern part of the city, had polluted the soil in Baia Mare area with Pb, Cd, Cu, Zn, and As [15–17].

In this study, concentration of Cu, Zn, Pb, Cd, Ni, Co, As, Cr and Hg in vineyard soil, several parts of grapevine (*Vitis vinifera* L.), as well as in must and wine from Baia Mare, Baia Sprie, and Simleul Silvaniei areas were analyzed.

2. Results and Discussion

2.1. Metal Concentration in Soil Samples

Elemental concentration varied among soil samples but were considerably higher than concentrations allowed by the Romanian Regulation of allowable quantities of hazardous and harmful substance in soil (Order of the Ministry of Waters, Forests and Environmental Protection No. 756/3 November 1997), as well as by the Council Directive 86/278/EEC for Protection of the Environment (European Communities Council 1986) (Table 1). Physical properties of soil samples are provided in Supplementary Table S1.

Regardless of sampling depth, the highest concentrations of Cu were recorded in Baia Sprie area, followed by the Baia Mare area and Simleul Silvaniei area (Table 1). In all cases, the concentrations significantly exceeded the normal values set by the corresponding legislation (20 mg/kg). These high concentrations can be attributed to the pollution factor (in Baia Mare and Baia Sprie areas) or the extensive usage of Cu-based plant protection products (in the Simleul Silvaniei area). Detected values were higher than those reported previously from this area (640.6 mg/kg) [17], (599.75 mg/kg) [18], (314.00 mg/kg) [16] or other wine-producing areas in Southeast Romania [19], but were within the range established for Copșa Mică (77–7675 mg/kg) [3].

Table 1. The content of heavy metals in soil from areas studied (mg/kg DW) (Mean ± standard deviation) ($n = 3$).

Areas	Depth (cm)	Cu M.P.L.**	Zn M.P.L.	Pb M.P.L.	Cd M.P.L.	Ni M.P.L.	Co M.P.L.	As M.P.L.	Cr M.P.L.	Hg M.P.L.
Normal Values		20 mg/kg	100 mg/kg	20 mg/kg	1 mg/kg	20 mg/kg	15 mg/kg	5 mg/kg	30 mg/kg	0.1 mg/kg
Alert threshold	Susceptible	100 mg/kg	300 mg/kg	50 mg/kg	3 mg/kg	75 mg/kg	30 mg/kg	15 mg/kg	100 mg/kg	1 mg/kg
	Less Susceptible	250 mg/kg	700 mg/kg	250 mg/kg	5 mg/kg	200 mg/kg	100 mg/kg	25 mg/kg	300 mg/kg	4 mg/kg
Intervention threshold	Susceptible	200 mg/kg	600 mg/kg	100 mg/kg	5 mg/kg	150 mg/kg	50 mg/kg	25 mg/kg	300 mg/kg	2 mg/kg
	Less Susceptible	500 mg/kg	1.500 mg/kg	1.000 mg/kg	10 mg/kg	500 mg/kg	250 mg/kg	50 mg/kg	600 mg/kg	10 mg/kg
Baia Mare	0–20	2510.52 ± 164.99 e γ	1637.98 ± 141.78 fγ	3074.29 ± 201.65 d β	14.13 ± 1.36 d β	28.60 ± 3.51 aα	29.57 ± 1.65 aα	4.13 ± 0.52 abcα	2.25 ± 0.79 ab αβ	0.058 ± 0.025 a α
	20–40	3317.02 ± 156.30 d β	1317.48 ± 68.87 gδ	3419.25 ± 196.78 c αβ	13.79 ± 0.74 d β	27.59 ± 1.28 aβ	19.91 ± 1.76 c β	3.16 ± 0.99 abcde αβ	1.49 ± 0.60 abcd β	0.053 ± 0.025 a α
	40–60	3146.25 ± 124.62 d β	2266.07 ± 93.68 e β	3118.06 ± 149.57 d β	15.66 ± 0.71 d β	25.43 ± 2.70 aβ	20.21 ± 1.09 bc β	4.05 ± 0.45 abcdα	2.72 ± 0.65 a α	0.053 ± 0.012 a α
	60–80	3687.25 ± 81.82 c α	2734.93 ± 147.45 d α	3554.34 ± 166.99 bc α	19.78 ± 1.41 c α	19.52 ± 0.79 b β	21.98 ± 1.69 b β	2.49 ± 0.57 bcde β	2.15 ± 0.01 abc αβ	0.042 ± 0.024 a α
	Average	3165.26 ± 165.26	1989.12 ± 112.92	3288.98 ± 178.75	15.84 ± 1.36	25.29 ± 2.07	22.57 ± 1.65	3.46 ± 0.63	2.15 ± 0.51	0.052 ± 0.021
Baia Sprie	0–20	4073.87 ± 182.03 aα	3134.45 ± 137.89 b βγ	3677.95 ± 148.11 b β	23.25 ± 1.25 β	18.00 ± 1.38 b α	8.41 ± 0.95 efg β	5.13 ± 1.22 a α	2.58 ± 1.25 ab α	0.068 ± 0.029 a α
	20–40	3998.09 ± 9.69 ab αβ	2934.62 ± 243.58 c γ	4262.23 ± 156.00 aα	19.52 ± 1.01 c γ	16.16 ± 1.91 b α	11.39 ± 1.03 d α	4.49 ± 2.49 ab α	2.35 ± 0.34 ab α	0.070 ± 0.017 a α
	40–60	3855.49 ± 58.38 bc β	3323.19 ± 157.94 ab αβ	4181.79 ± 144.30 d β	32.53 ± 0.99 aα	17.22 ± 2.33 b α	7.18 ± 0.99 fgh βγ	4.72 ± 1.01 a α	2.39 ± 1.15 ab α	0.067 ± 0.012 a α
	60–80	4155.95 ± 79.30 aα	3483.25 ± 94.11 aα	4127.23 ± 193.63 aα	32.06 ± 1.40 aα	18.23 ± 0.40 b α	6.01 ± 1.33 gh γ	3.49 ± 1.52 abcd α	1.94 ± 0.37 abcd α	0.052 ± 0.008 a α
	Average	4020.85 ± 86.85	3218.88 ± 158.88	4062.30 ± 147.01	26.84 ± 1.25	17.40 ± 1.51	8.25 ± 1.08	4.46 ± 1.56	2.31 ± 0.78	0.064 ± 0.016
Simleul Silvaniei	0–20	621.79 ± 64.64 gβ	76.86 ± 7.71 h α	12.62 ± 2.76 e α	0.27 ± 0.04 e α	7.82 ± 1.81 d β	5.08 ± 1.77 h β	2.04 ± 0.03 de α	0.96 ± 0.41 cd α	0.048 ± 0.032 a α
	20–40	791.71 ± 50.85 fg α	68.18 ± 3.09 h α	6.62 ± 0.45 e β	0.22 ± 0.15 e α	12.14 ± 1.53 c α	8.80 ± 1.05 ef α	2.16 ± 0.03 cde α	0.68 ± 0.53 cd α	0.060 ± 0.017 a α
	40–60	842.88 ± 68.11 fα	45.36 ± 10.19 h β	6.95 ± 1.68 e β	0.15 ± 0.06 e α	5.89 ± 1.17 d β	10.08 ± 1.24 de α	2.05 ± 0.57 de α	1.39 ± 0.21 bcd α	0.57 ± 0.006 a α
	60–80	793.69 ± 8.64 fg α	45.56 ± 9.79 h β	7.76 ± 1.83 e β	0.12 ± 0.10 e α	6.97 ± 0.49 d β	8.43 ± 0.93 efg α	1.15 ± 0.01 e β	1.45 ± 0.61 abcd α	0.034 ± 0.010 a α
	Average	762.52 ± 48.06	59.99 ± 6.42	8.49 ± 1.66	0.19 ± 0.09	8.20 ± 1.25	8.10 ± 1.24	1.81 ± 0.33	1.21 ± 0.44	0.050 ± 0.016
Average		2649.54 ± 88.95	1755.66 ± 93.00	2453.26 ± 109.14	14.29 ± 0.90	16.96 ± 1.61	12.97 ± 1.32	3.24 ± 0.84	1.77 ± 0.50	0.055 ± 0.066
Minimum values		621.79 ± 64.64	45.36 ± 10.19	6.62 ± 0.45	0.12 ± 0.10	6.97 ± 0.49	5.08 ± 1.77	1.15 ± 0.01	0.68 ± 0.53	0.034 ± 0.010
Maximum values		4155.95 ± 79.30	3483.25 ± 94.11	4262.23 ± 156.00	32.53 ± 0.99	28.60 ± 3.51	29.57 ± 1.65	5.13 ± 1.22	2.72 ± 0.65	0.070 ± 0.017
Sig.		***	***	***	***	***	***	**	*	in

Table 1. *Cont.*

Areas	Depth (cm)	Cu M.P.L.**	Zn M.P.L.	Pb M.P.L.	Cd M.P.L.	Ni M.P.L.	Co M.P.L.	As M.P.L.	Cr M.P.L.	Hg M.P.L.
Huzum et al., 2012 [20]		256.00	60.10	12.90	0.21	29.90	7.20	11.20	11.04 ± 0.78	–
Albulescu et al., 2009 [13]		–	–	21.90	1.77	24.55	–	–	13.32	–
Alagić et al., 2015 [21]		293.00	–	42.80	3.14 ± 0.03	16.67 ± 0.09	–	10.70 ± 0.01	–	–
Bravo et al., 2017 [22]		10.87 ± 5.10	–	16.18 ± 5.20	–	–	–	–	–	–
Bora et al., 2015 [23]		479.64 ± 53.97	69.44 ± 4.02	14.77 ± 0.74	0.45 ± 0.10	16.28 ± 2.01	9.75 ± 1.47	–	–	–
Bora et al., 2018 [24]		356.03 ± 4.36		7.00 ± 0.81	0.37 ± 0.05	5.68 ± 0.53	3.73 ± 0.48	1.29 ± 0.10	11.04 ± 0.78	0.075 ± 0.013
European Communities Council 1986		50–140	150–300	50–300	1–3	30–75	–	–	–	1–1.5
Common abundance in topsoil's [c]		5–50	10–100	–	0.1–1	20–50	–	0.1–55	–	–
Kabata-Pendias, 2010 [25]		13–24	45–100	22–44	0.37–0.78	12.0–34	–	0–9.3	–	–
Phytotoxic levels of elements in soils [c]		36–698	100–1,000	–	–	100	–	200	–	–

Average value ± standard deviation ($n = 3$). Greek letters are significance of difference ($p \leq 0.005$) for the same type of soil but different profile (depth). Roman letters are significance of difference ($p \leq 0.05$) between the depths of the soil profile. The difference between any two values, followed by at least one common letter, is insignificant. *Order of the Ministry of Waters, Forests and Environmental Protection No. 756/3 November 1997, approving the regulation on the assessment of environmental pollution, Bucharest, Romania; 1997. **M.A.L. (Maximum Admissible Limit) = Normal Values. in = insignificant

The results obtained by Damian et al. (44–5823 mg/kg) are comparable to those obtained in this research. The Cu values obtained for Simleul Silvaniei are conformable with those recorded by Alagić et al. (293.00 mg/kg) [21] and Bora et al. (479.64 mg/kg) [23].

In the Baia Sprie and Baia Mare areas, the concentrations of Zn tended to increase with the sampling depth, with the highest concentrations being detected in samples collected at 60–80 cm (3483.25 ± 94.11 mg/kg and 2734.93 ± 147.45 mg/kg, respectively). All values greatly exceeded the normal levels of Zn allowed by the law (100 mg/kg). In contrast, in the Simleul Silvaniei area, the highest concentration was recorded in the surface soil profile (76.86 ± 7.71 mg/kg (0–20 m)), and it tended to decrease with the increasing soil depth. The Zn values obtained are higher than data published in previous reports from Baia Mare or other Romanian regions [16,26,27].

Concentrations of Pb and Cd varied within a wide range (Table 1). The highest concentrations were recorded in the Baia Mare and Baia Sprie areas, significantly higher than those detected in Simleul Silvaniei or allowed by the applicable legislation (20 mg/kg for Pb, 1 mg/kg for Cd). The extremely high values of Pb and Cd indicate severe heavy metal pollution in these two areas. Similar [18] or lower [16,27] values were also reported from these regions. The average content of Ni and Co in soil samples exceeded the normal concentration in the Baia Mare area (25.29 ± 2.07 vs 20 mg/kg for Ni and 22.57 ± 1.65 vs 15 mg/kg for Co), but were below the limit in the other two areas (Table 1). For comparison, Mihali et al. recorded similar values (13.1 mg/kg [Ni] and 24.8 mg/kg [Co]) in the Baia Mare area [19], while another study conducted in unpolluted regions from Dobrogea and Muntenia reported Ni concentrations between 0.97–11.29 mg/kg and Co concentrations between 0.49–4.36 mg/kg [26].

The concentrations of As, Cr, and Hg indicated no pollution of the soil samples with these heavy metals; values were below the normal levels. The highest values were obtained in Baia Sprie (4.46 ± 1.56 mg/kg As; 2.31 ± 0.78 mg/kg Cr; 0.064 ± 0.016 mg/kg Hg) followed by Baia Mare (3.46 ± 0.63 mg/kg As; 2.51 ± 0.51 mg/kg Cr; 0.052 ± 0.021 mg/kg Hg) area. A recent study conducted in Vaslui county reported higher content of As (10.14 mg/kg) and Cr (62.05 mg/kg) compared to our results [28].

2.2. Metal Concentration in Plant Material Samples

2.2.1. Metal Concentration in Roots

Roots are in direct contact with the soil solution and the concentration of heavy metals in roots is generally used as indicative of soil metal bioavailability [29]. Varieties cultivated in Simleul Silvaniei showed the lowest concentrations of Cu and Zn, compared to the varieties from Baia Mare and Baia Sprie. Italian Riesling from Baia Mare and Baia Sprie (779.15 ± 4.66 mg/kg and 670.51 ± 6.61 mg/kg, respectively) and Feteasca alba from Baia Sprie (669.15 ± 21.27 mg/kg) contained the highest concentration of Cu while the varieties cultivated in Baia Mare area recorded the highest Zn concentration (Table 2). Studies have shown that high concentration of Cu can affect the growth of the roots [9,30]. The highest concentrations of Pb was registered in Feteasca regala from Baia Mare (60.81 ± 5.95 mg/kg), in Italian Riesling from Baia Sprie (92.26 ± 1.11 mg/kg), significantly higher as compared to the same varieties grown in the Simleul Silvaniei area (0.83 ± 0.60 mg/kg (Feteasca regala) and 0.43 ± 0.17 mg/kg (Italian Riesling)). According to Vamerali et al., Pb has no important role in functions of plants [7]. Roots of Feteasca regala from Baia Mare and Feteasca alba from Baia Sprie had the highest concentration of Cd (7.09 ± 0.83 mg/kg and 3.07 ± 0.12 mg/kg, respectively) and Co (32.24 ± 1.23 mg/kg and 10.95 ± 1.26 mg/kg) (Table 2). Cd and Co concentrations in these areas were significantly higher than those obtained in varieties grown from Simleul Silvaniei. In Baia Sprie and Simleul Silvaniei, concentrations of Ni and As were similar amongst varieties, while in Baia Mare area, Italian Riesling variety had higher concentration of As compared to Feteasca alba and Feteasca regala (Table 2). Interestingly, roots from Simleul Silvaniei showed higher content of Ni compared to varieties from Baia Sprie. Concentration of Cr was similar for all three varieties in Baia Mare and Baia Sprie, except for Feteasca regala from Baia Mare which recorded a significantly higher concentration.

Table 2. The content of heavy metals in plant samples (mg/kg DW)(Mean ± standard deviation) (n = 3).

Areas	Variety	Plant Parts	Cu	Zn	Pb	Cd	Ni	Co	As	Cr	Hg	Sig
Baia Mare	Feteasca alba	Roots	450.31 ± 13.17 d α	189.80 ± 1.19 c α	49.61 ± 6.76 g γ	4.27 ± 0.22 e α	11.66 ± 1.62 e β	19.56 ± 2.48 e β	1.11 ± 0.22 ghijk β	2.54 ± 0.39 c α	0.025 ± 0.006 defg β	***
		Canes	72.03 ± 2.75 g β	105.68 ± 5.57 ij	64.63 ± 5.17 ef β	4.49 ± 0.29 e α	8.43 ± 0.85 gh γ	40.31 ± 2.23 c α	1.33 ± 0.33 fghi β	0.67 ± 0.09 e β	0.024 ± 0.009 defg β	***
		Leaves	61.65 ± 1.71 g β	119.47 ± 1.58 hi β	89.64 ± 1.87 bc α	2.39 ± 0.02 g β	25.25 ± 1.14 a α	6.89 ± 1.01 fgh γ	3.13 ± 0.68 a α	0.40 ± 0.18 e β γ	0.049 ± 0.008 b α	***
		Grapes	8.49 ± 0.64 jk γ	9.60 ± 0.98 kl δ	6.34 ± 1.06 klm δ	0.69 ± 0.05 jkl γ	0.74 ± 0.25 o δ	1.26 ± 0.65 h δ	0.60 ± 0.30 ijkl β	0.13 ± 0.01 e δ	0.013 ± 0.002 ijk β	***
		Average	148.12 ± 4.57	106.14 ± 2.33	52.56 ± 3.70	2.96 ± 0.15	11.52 ± 0.97	17.01 ± 1.59	1.54 ± 0.38	0.94 ± 0.17	0.028 ± 0.006	
	Feteasca regala	Roots	460.00 ± 4.00 α	211.18 ± 4.81 ab α	60.81 ± 5.95 f β	7.09 ± 0.83 c α	12.12 ± 1.97 e β	32.24 ± 1.23 d α	0.90 ± 0.60 ghijkl β	3.48 ± 0.24 a α	0.016 ± 0.005 fghijk γ	***
		Canes	77.31 ± 3.76 g	119.65 ± 5.75 hi β	68.11 ± 3.76 de α	5.86 ± 0.23 d β	6.46 ± 0.54 hijk γ	33.86 ± 1.92 d α	1.20 ± 0.44 fghi β	0.53 ± 0.23 e β	0.027 ± 0.002 def β	***
		Leaves	65.66 ± 1.88 g γ	114.52 ± 2.33 hi β	73.47 ± 2.64 d α	1.99 ± 0.02 gh γ	22.78 ± 0.82 bc α	6.17 ± 0.50 gh β	2.17 ± 0.50 bcd α	0.42 ± 0.18 e β γ	0.049 ± 0.009 b α	***
		Grapes	12.30 ± 2.39 hijkδ	10.13 ± 1.33 kl γ	6.19 ± 0.95 klm γ	0.58 ± 0.03 jkl δ	0.77 ± 0.11 o δ	1.08 ± 0.62 γ	0.36 ± 0.37 kl β	0.13 ± 0.03 e γ	0.011 ± 0.003 jk γ	***
		Average	153.82 ± 3.01	113.87 ± 3.56	52.15 ± 3.33	3.88 ± 0.28	10.53 ± 0.86	18.34 ± 1.07	1.16 ± 0.48	1.14 ± 0.17	0.026 ± 0.005	
	Italian Riesling	Roots	779.15 ± 4.66 a	174.58 ± 3.70 d α	32.07 ± 1.76 i γ	6.42 ± 0.25 cd α	11.56 ± 1.40 e α	23.37 ± 1.40 e α	2.31 ± 0.76 bcd α	2.64 ± 1.23 bc α	0.011 ± 0.003 jk β	***
		Canes	72.93 ± 2.25 g β	99.45 ± 4.03 j γ	40.23 ± 4.24 h β	4.64 ± 0.13 e β	6.82 ± 1.28 ghij γ	12.42 ± 1.62 f β	2.09 ± 0.13 bcde α	0.58 ± 0.15 e β	0.022 ± 0.007 defghij α	***
		Leaves	76.82 ± 1.92 g β	127.58 ± 1.55 gh β	90.89 ± 1.55 bc α	2.11 ± 0.03 g γ	21.54 ± 0.88 c α	4.84 ± 0.88 gh γ	1.87 ± 0.80 def α	0.19 ± 0.11 e β	0.029 ± 0.006 cde α	***
		Grapes	9.66 ± 1.02 ijk γ	8.52 ± 1.25 kl δ	4.60 ± 0.64 klm δ	0.75 ± 0.15 jkl δ	0.60 ± 0.17 o δ	0.91 ± 0.06 h δ	0.29 ± 0.22 l β	0.14 ± 0.05 e β	0.012 ± 0.003 ijk β	***
		Average	234.64 ± 2.46	102.53 ± 2.63	41.95 ± 2.05	3.48 ± 0.14	10.13 ± 0.93	10.39 ± 0.98	1.64 ± 0.48	0.89 ± 0.039	0.019 ± 0.005	
Baia Sprie	Feteasca alba	Roots	661.74 ± 14.49 b α	160.93 ± 3.58 de β	84.87 ± 0.65 c β	3.07 ± 0.12 f γ	3.42 ± 0.90 lmn β	10.95 ± 1.26 fg β	1.80 ± 0.62 defg β	2.38 ± 0.28 cd α	0.020 ± 0.002 efghijk β γ	***
		Canes	124.56 ± 9.02 f γ	192.83 ± 16.12 c α	85.62 ± 12.97 c α	9.50 ± 0.83 a α	3.41 ± 0.38 lmn β	85.37 ± 7.79 a α	0.92 ± 0.20 hijkl γ	0.61 ± 0.12 e β	0.026 ± 0.010 def β	***
		Leaves	148.02 ± 16.94 e β	163.54 ± 26.93 d αβ	96.55 ± 9.07 bc α	6.50 ± 0.91 cd β	22.29 ± 1.60 bc α	5.31 ± 0.74 gh β	2.77 ± 0.18 ab α	0.25 ± 0.04 e γ	0.053 ± 0.003 b α	***
		Grapes	12.31 ± 1.82 hijkδ	8.91 ± 1.50 kl γ	8.91 ± 1.50 jklm γ	1.22 ± 0.46 ij δ	1.68 ± 0.69 no β	3.26 ± 0.69 h β	0.90 ± 0.07 hijkl γ	0.13 ± 0.02 e γ	0.014 ± 0.002 hijk γ	***
		Average	236.66 ± 10.57	131.55 ± 12.03	68.98 ± 6.05	5.07 ± 0.58	7.70 ± 0.89	26.22 ± 2.62	1.60 ± 0.27	0.84 ± 0.12	0.028 ± 0.004	

Table 2. Cont.

Areas	Variety	Plant Parts	Cu	Zn	Pb	Cd	Ni	Co	As	Cr	Hg	Sig
Simleul Silvaniei	Feteasca regala	Roots	670.51 ± 6.61 b α	115.61 ± 9.52 hi β	72.87 ± 11.00 d β	1.82 ± 0.44 ghi β	2.85 ± 0.30 mno β	4.78 ± 0.19 gh β	2.65 ± 0.96 abc α	2.89 ± 0.69 b α	0.020 ± 0.003 efghijk γ	***
		Canes	143.72 ± 2.46 e β	198.05 ± 11.47 bc α	93.63 ± 12.77 bc β	9.60 ± 0.95 a α	6.99 ± 0.84 ghij β	76.96 ± 13.65 b α	0.81 ± 0.17 hijkl β	0.47 ± 0.12 e β	0.029 ± 0.003 cde β	***
		Leaves	143.83 ± 40.38 e β	134.99 ± 18.42 fg β	95.52 ± 7.57 b α	9.43 ± 0.80 a α	24.16 ± 4.50 ab α	4.96 ± 1.06 gh β	2.50 ± 0.08 abcd α	0.32 ± 0.11 e β	0.054 ± 0.005 b α	***
		Grapes	13.75 ± 1.17 hijk γ	6.80 ± 2.47 kl γ	6.80 ± 2.47 klm γ	1.13 ± 0.53 ij β	2.56 ± 1.60 mno β	3.90 ± 1.84 h β	0.86 ± 0.08 hijkl β	0.11 ± 0.03 e β	0.012 ± 0.003 ijk δ	***
		Average	242.95 ± 12.66	113.86 ± 10.47	67.21 ± 8.45	5.50 ± 0.68	9.14 ± 1.81	22.65 ± 4.19	0.71 ± 0.32	0.95 ± 0.24	0.029 ± 0.004	
	Italian Riesling	Roots	669.15 ± 21.27 b α	97.66 ± 1.14 j γ	92.26 ± 1.11 bc β	1.27 ± 0.29 hij γ	2.31 ± 0.88 mno γ	6.13 ± 1.25 gh β	2.53 ± 0.40 abcd α	2.05 ± 0.56 d α	0.014 ± 0.002 hij γ	***
		Canes	147.92 ± 21.61 e β	221.13 ± 6.57 a α	102.91 ± 0.57 a α	8.66 ± 0.60 b α	8.14 ± 1.69 ghi β	76.93 ± 10.85 b α	0.80 ± 0.20 hijkl β	0.55 ± 0.10 e β	0.025 ± 0.001 defg β	***
		Leaves	130.98 ± 22.64 ef β	147.63 ± 30.72 ef β	106.32 ± 14.48 a α	6.69 ± 0.38 c β	18.69 ± 1.95 d α	6.21 ± 1.36 gh β	1.92 ± 0.80 cdef α	0.16 ± 0.06 e β	0.067 ± 0.009 a α	***
		Grapes	14.51 ± 1.25 hijk γ	6.78 ± 2.14 kl δ	6.74 ± 1.22 klm γ	0.88 ± 0.10 jk γ	1.13 ± 0.40 no γ	4.60 ± 2.54 gh β	0.83 ± 0.17 hijkl β	0.12 ± 0.03 e β	0.013 ± 0.002 ijk γ	***
		Average	240.64 ± 16.69	118.32 ± 10.14	77.06 ± 4.35	4.38 ± 0.34	7.57 ± 1.23	23.47 ± 4.00	1.52 ± 0.39	0.72 ± 0.19	0.030 ± 0.004	–
	Feteasca alba	Roots	10.01 ± 0.39 hijk γ	6.00 ± 1.49 kl γ	1.18 ± 0.11 m β	0.67 ± 0.17 jkl β	5.41 ± 0.74 jkl β	2.38 ± 0.31 h α	1.23 ± 0.22 fghi β	0.42 ± 0.24 e α	0.017 ± 0.005 fghijk β	***
		Canes	30.46 ± 2.70 h α	18.66 ± 0.88 k	1.21 ± 0.14 m α	0.08 ± 0.04 kl γ	4.29 ± 0.92 klm β	3.02 ± 0.81 h α	0.62 ± 0.02 hijkl γ	0.27 ± 0.07 e α β	0.022 ± 0.009 defghi β	***
		Leaves	26.10 ± 2.79 hij β	12.23 ± 1.72 kl β	1.67 ± 0.23 m β	0.11 ± 0.02 kl γ	9.01 ± 0.39 fg α	1.36 ± 0.07 h β	2.43 ± 0.18 abcd α	0.10 ± 0.01 e β	0.038 ± 0.007 c α	***
		Grapes	2.31 ± 0.77 k δ	1.25 ± 0.53 l δ	0.49 ± 0.34 m β	1.09 ± 0.03 ij α	1.11 ± 0.30 no γ	1.44 ± 0.29 h β	0.17 ± 0.07 l δ	0.09 ± 0.02 e β	0.011 ± 0.002 jk β	***
		Average	17.22 ± 1.66	9.54 ± 1.16	1.14 ± 0.21	0.49 ± 0.07	4.96 ± 0.59	2.05 ± 0.37	1.11 ± 0.12	0.22 ± 0.09	0.022 ± 0.006	–
	Feteasca regala	Roots	17.19 ± 3.42 hijk β	5.69 ± 0.56 kl γ	0.83 ± 0.60 m β	0.66 ± 0.44 jkl β	5.30 ± 1.08 jkl β	1.59 ± 0.29 h β	1.31 ± 0.14 fghi γ	0.49 ± 0.12 e α	0.015 ± 0.006 ghijk β	***
		Canes	26.98 ± 4.05 hij α	20.32 ± 0.70 α	1.62 ± 0.06 lm α	0.05 ± 0.031 γ	6.03 ± 0.50 ijk β	3.36 ± 0.33 h α	0.63 ± 0.05 hijkl γ	0.20 ± 0.06 e β	0.025 ± 0.004 defg α	***
		Leaves	31.41 ± 1.87 k α	16.13 ± 1.32 kl β	2.33 ± 0.29 lm α	0.13 ± 0.03 kl γ	10.69 ± 0.68 ef α	1.48 ± 0.14 h β	2.37 ± 0.27 bcd α	0.09 ± 0.04 e β	0.031 ± 0.007 cd α	***
		Grapes	3.38 ± 1.76 k γ	1.13 ± 0.50 l δ	0.40 ± 0.15 m β	1.30 ± 0.23 ij α	0.89 ± 0.12 o γ	1.48 ± 0.32 h β	0.23 ± 0.10 l δ	0.09 ± 0.00 e β	0.011 ± 0.002 k β	***
		Average	19.74 ± 2.78	10.82 ± 0.77	1.30 ± 0.28	0.54 ± 0.18	5.73 ± 0.60	1.98 ± 0.27	1.14 ± 0.14	0.22 ± 0.06	0.021 ± 0.005	–

Table 2. Cont.

Italian Riesling	Roots	9.47 ± 0.85 ijk β	5.28 ± 3.58 kl γ	0.43 ± 0.17 m β	1.19 ± 0.34 ij α	5.58 ± 0.58 jkl β	1.65 ± 0.44 h β	1.39 ± 0.46 efgh β	0.51 ± 0.06 e α	0.013 ± 0.002 ijk β	***
	Canes	27.70 ± 2.10 hij α	20.64 ± 1.41 k α	1.09 ± 0.03 m β	0.10 ± 0.02 kl β	6.60 ± 1.20 hijk β	2.75 ± 0.78 h α	0.43 ± 0.02 jkl γ	0.25 ± 0.11 e β	0.023 ± 0.002 defgh α	***
	Leaves	22.02 ± 2.18 hijk α	11.77 ± 1.11 kl β	1.89 ± 0.14 m α	0.14 ± 0.01 kl β	8.77 ± 1.04 fgh α	1.43 ± 0.03 h β	2.25 ± 0.34 bcd α	0.09 ± 0.06 e γ	0.027 ± 0.008 def α	***
	Grapes	2.85 ± 0.29 k γ	1.11 ± 0.621 δ	0.59 ± 0.28 m β	1.21 ± 0.67 ij α	1.46 ± 0.38 no γ	1.93 ± 0.09 h α β	0.13 ± 0.03 l γ	0.11 ± 0.02 e γ	0.013 ± 0.002 β	***
Average		15.51 ± 1.36	9.70 ± 1.68	1.00 ± 0.16	0.66 ± 0.26	5.60 ± 0.80	1.94 ± 0.34	1.05 ± 0.21	0.24 ± 0.06	0.019 ± 0.004	-
Average		148.43 ± 6.30	81.18 ± 5.05	41.27 ± 3.24	3.05 ± 0.30	8.15 ± 0.37	14.34 ± 1.81	1.37 ± 0.31	0.69 ± 0.16	0.026 ± 0.004	-
Sig.		***	***	***	***	***	***	***	***	***	-
Areas		***	***	***	***	***	***	***	***	***	-
Variety		***	***	in	***	in	***	in	*	*	-
Plant parts		***	***	***	***	***	***	***	***	***	-
Areas × Variety		***	***	***	***	***	***	*	in	**	-
Areas × Plant parts		***	***	***	***	***	***	***	***	***	-
Variety × Plant parts		***	***	***	*	***	***	***	*	in	-
Areas × Variety × Plant part		***	***	***	***	*	***	in	in	*	-
Normal range in plant tissues		4–15 a,b	60 b 8–100 d	0.1–10 c 1–13 d	0.1–2.4 d	0.05–10 b 1 d	-	0.009–1.5 b,c	-	-	-
Phytotoxic concentration in plant tissues		15–20 a,c,e 4–40 for leaves and 100–400 for root b	100–500 b	10–20 a	5–10 a,c,e	20–30 a	-	>20 a	-	-	-
			150–200 a,c,e			>10 b,e	-	1–20 c 10–100 e	-	-	-

Average value ± standard deviation (n = 3). Greek letters are significance of difference ($p \leq 0.005$) for the same type of soil but different profile (depth). Roman letters are significance of difference ($p \leq 0.05$) between the plant parts of the same variety. The difference between any two values, followed by at least one common letter, is insignificant. in = insignificant.
[a] Vamerali et al., 2010 [7]; [b] Alloway, 2013 [31]; [c] Kabata-Pendias, 2010 [25].

The varieties grown in Simleul Silvaniei have a lower concentration compared to varieties grown in Baia Mare and Baia Sprie. Hg was detected in low concentrations in all three areas. The observed concentrations of Cu, Zn, Ni, and As exceed the toxic threshold in plant tissues [7,31]. Overall, data suggests that high concentrations of heavy metals in soil result an increased metal content in the roots as well. Compared to our findings, grapevines grown in polluted areas from East Serbia have shown similar concentrations of heavy metals in roots [21].

2.2.2. Metal Concentration in Canes

Cu has the highest concentrations in all varieties cultivated in Baia Sprie (147.92 ± 2.46 mg/kg (Italian Riesling); 143.72 ± 2.46 mg/kg (Feteasca Regala) 124.56 ± 9.02 mg/kg (Feteasca Alba), followed by the Baia Mare area (77.31 ± 3.76 mg/kg (Feteasca Regala); 72.93 ± 2.15 mg/kg (Italian Riesling); 72.03 ± 2.75 (Feteasca Alba). This can be explained with the heavy metal pollution phenomenon. Though Cu is involved in many vital processes in plants such as photosynthesis, flowering, seed production, and plant growth, its excessive concentrations may cause a significant modification of biochemical processes, leading to the reduction of shoot growth [7,32]. Results obtained in Baia Mare are comparable with those reported from Turulung, NW Romania (63.67 ± 2.67 mg/kg) [32] and much lower than those obtained from polluted regions from East Serbia (170.90 ± 0.80 mg/kg Cu [Flotacijsko Jalovište]; 175.00 ± 2.00 mg/kg [Bolničko naselje]; 160.00 ± 0.90 mg/kg [Slatinsko naselje]) [21]. Analyzing the concentration of Zn in the canes, varieties cultivated in Baia Sprie had the highest concentrations. Cane samples from the Baia Mare area also displayed high concentrations of Zn, exceeding the toxicity threshold in plant tissues [7,31]. The lower concentrations detected in the Simleul Silvaniei area were consistent with the literature values reported for other areas [21,23]. Regarding the concentration of Pb, Cd, Ni, and Co in the string and canes, values recorded in the Baia Sprie and Baia Mare areas are significantly higher than those recorded for the same heavy metals in the Simleul Silvaniei area (Table 2) or reported from other regions [21,23]. In all regions and varieties studied, concentrations of As, Cr and Hg were similar and below the toxicity threshold in plant tissues.

2.2.3. Metal Concentration in Leaves

Agricultural crops are especially sensitive to Cu concentration. As a first signal of excessive supply of Cu, symptoms of chlorosis may occur [32]. In this study, significantly higher concentrations of Cu were detected in Baia Sprie as compared to Baia Mare. No significant differences in Cu concentrations were found between varieties cultivated in the same area, except for Simleul Silvaniei, where leaves of Feteasca regala had higher Cu content as the other two varieties tested. Similarly, Zn concentrations were highest in Baia Sprie. Feteasca alba leaves from Baia Sprie and Italian Riesling leaves from Baia Mare had significantly higher content of Zn than leaves of other varieties collected from the same area. For both Cu and Zn, concentrations were above the phytotoxic threshold. Concentrations of Pb, Cd, and Ni in leaves were similar across varieties cultivated in the same area. While Cd and Ni did not exceed the phytotoxic concentrations established for plant tissues, the concentrations of Pb in leaves collected from Baia Sprie and Baia Mare areas were greatly above the pre-defined phytotoxic concentration, which can be attributed to the pollution factor in these areas. Concentrations of Co were similar in Baia Mare and Baia Sprie areas ranging between 4.84 ± 0.88 mg/kg and 6.89 ± 1.01 mg/kg (Table 2). The levels in leaves were slightly above the normal range [25], but still below the phytotoxic concentration.

2.2.4. Metal Concentration in Grapes

According to Vamerali et al., Cu is a constituent of enzymes involved in photosynthesis, in reproductive phase, and in determining the yield and quality in crops. Zn is a constituent of cell membranes and it is involved in DNA transcription, activation of enzymes, and evaluation of the yield and quality of crops [7]. Varieties cultivated in Baia Mare and Baia Sprie areas recorded comparable concentrations of Cu (8.49 ± 0.64–12.30 ± 2.39 mg/kg and 12.31 ± 1.82–14.51 ± 1.25 mg/kg, respectively) and Zn (8.52 ± 1.25–10.13 ± 1.33 mg/kg and 6.78 ± 2.14–8.91 ± 1.50 mg/kg), but values were comparable and within the normal range accepted in plant tissues (Table 2). While these concentrations can be attributed to the heavy metal pollution phenomenon in the two areas, Cu and Zn content of varieties cultivated in Simleul Silvaniei area (2.31 ± 0.77–3.38 ± 1.76 mg/kg and 1.11 ± 0.62–1.25 ± 0.53 mg/kg, respectively) can be ascribed to plant protection products or vine nutrition process. Values of the Cu and Zn concentration are higher than those reported in Brazil (79.87 ± 0.05 μg/100g grape berries - Cabernet Sauvignon and 31.56 ± 0.04 μg/100g grape berries - Merlot for Cu; 42.47 ± 0.17 μg/100g and 52.24 ± 0.74μg/100g for Zn) [33].

Although Pb occurs naturally in all plants, it has not been shown to play any essential role in their metabolism and its concentration at the level of 2–6 μg/g should be sufficient [25]. Pb has recently received much attention as a major metallic pollutant of the environment and as an element toxic to plants. Feteasca alba variety cultivated in Baia Sprie showed the highest Pb content (8.91 ± 1.50 mg/kg), other varieties from Baia Sprie and Baia Mare areas having similar Pb concentration (between 4.60 ± 0.64 and 6.80 ± 2.47 mg/kg). The varieties grown in Simleul Silvaniei recorded significantly lower concentration of Pb in grapes (Table 2). Overall, concentrations of Cd and Ni were detected in similar ranges in all three areas, though values tended to be higher in Baia Sprie and Simleul Silvaniei regions compared to Baia Mare. Cd is considered a non-essential element for metabolic processes; it is effectively absorbed by root and leaf systems and is also accumulated in soil organisms. There are evidences that an appreciable fraction of Cd is taken up passively by roots, but Cd is also absorbed metabolically [25]. There is no evidence of an essential role of Ni in plant metabolism, although several investigators suggested that Ni might be essential for plants. The essentiality of Ni for some biosynthesis of a number of bacteria has been proven. Also, its role in the nodulation of legumes and effects on the nitrification and mineralization of some OM was described [25]. Concentration of Co was higher in varieties from Baia Sprie area (3.26 ± 0.69–4.60 ± 2.54 mg/kg) as compared to Baia Mare (0.91 ± 0.06–1.26 ± 0.65 mg/kg) and Simleul Silvaniei (1.44 ± 0.29–1.93 ± 0.09 mg/kg). Co is cofactor of biosynthetic enzymatic activities essential for *Rhizobium*. Its content in plants is highly controlled by both soil factors and the ability of plants to absorb this metal [34]. In higher plants, absorption of Co by roots involves active transport [25]. Varieties from Baia Sprie had the highest As concentrations (0.83 ± 0.17–0.90 ± 0.07 mg/kg), followed by varieties from Baia Mare (0.29 ± 0.22–0.60 ± 0.30 mg/kg). No significant differences in Cr and Hg content were observed in all grape samples. The biochemistry of Hg is associated mainly with biological transformation of its compounds. However, it is not clear yet which processes are the most important in its cycling in the environment. In general, Hg content of plants is high when the Hg content of soils is also high, but this relation does not always hold. The results obtained are much higher than those reported in other studies [21,23,25,31].

2.3. Metal Concentration in Must and Wine

2.3.1. Metal Concentration in must

Concentrations of Cu and Zn in must samples from Baia Sprie and Baia Mare areas exceeded the maximum permissible limit (M.P.L.) (10 mg/L), indicating a serious Cu and Zn pollution of the corresponding areas. Concentrations in the varieties cultivated in Simleul Silvaniei were below this threshold (Table 3). In grapevine nutrition, small quantities of Zn are taken from the soil (Zn is a trace mineral), so it is naturally present in must and wine. During alcoholic fermentation, part of the Zn precipitates due to the reducing environment and is accumulated in yeast.

Table 3. The content of metal concentration in must and wine samples (Mean ± standard deviation) (n = 3).

Areas	Variety	Sample	Cu mg/L M.P.L. 1 mg/L	Zn mg/L M.P.L. 5 mg/L	Pb mg/L M.P.L. 0.15 mg/L	Cd mg/L M.P.L. 0.01 mg/L	Ni mg/L M.P.L. -	Co µg/L M.P.L. -	As µg/L M.P.L. 0.2 mg/L	Cr µg/L M.P.L. -	Hg µg/L M.P.L. -	Sig
Baia Mare	Feteasca alba	Must	24.87 ± 1.77 c	12.76 ± 2.19 cd	0.36 ± 0.03 c	0.05 ± 0.01 b	1.18 ± 0.06 a	LOQ	33.06 ± 1.58 bc	634.14 ± 6.44 d	0.20 ± 0.04 abc	***
		Wine	1.47 ± 0.09 g	5.59 ± 0.12 fg	0.17 ± 0.03 cd	0.04 ± 0.03 b	0.05 ± 0.01 f	LOQ	30.40 ± 1.96 cd	652.56 ± 5.56 c	0.11 ± 0.02 e	***
	Feteasca regala	Must	20.64 ± 0.90 d	11.59 ± 2.84 cde	0.69 ± 0.05 b	0.05 ± 0.02 b	0.59 ± 0.12 b	LOQ	48.30 ± 1.27 a	642.24 ± 9.54 cd	0.18 ± 0.03 bcd	***
		Wine	1.13 ± 0.04 g	5.36 ± 0.08 fg	0.27 ± 0.02 cd	0.03 ± 0.01 b	0.07 ± 0.03 f	LOQ	37.02 ± 2.23 b	646.26 ± 4.54 cd	LOQ f	***
	Italian Riesling	Must	20.36 ± 2.81 d	14.04 ± 1.93 c	0.34 ± 0.03 cd	0.03 ± 0.02 b	0.57 ± 0.18 b	LOQ	46.35 ± 2.60 a	548.50 ± 2.37 e	0.15 ± 0.05 cde	***
		Wine	1.20 ± 0.06 g	5.80 ± 0.11 fg	0.20 ± 0.03 cd	0.02 ± 0.01 b	0.04 ± 0.02 f	LOQ	34.17 ± 1.07 bc	645.06 ± 7.58 cd	LOQ f	***
Baia Sprie	Feteasca alba	Must	32.52 ± 3.26 a	25.83 ± 3.01 a	1.14 ± 0.49 a	0.21 ± 0.11 a	0.31 ± 0.03 de	LOQ	35.68 ± 3.29 b	431.67 ± 10.03 g	0.23 ± 0.02 ab	***
		Wine	2.46 ± 1.13 g	6.04 ± 1.70 fg	0.35 ± 0.14 c	0.06 ± 0.03 b	0.08 ± 0.02 f	LOQ	26.52 ± 3.61 de	317.81 ± 11.72 i	LOQ	***
	Feteasca regala	Must	29.27 ± 2.83 b	28.50 ± 7.86 a	1.13 ± 0.14 a	0.18 ± 0.04 a	0.42 ± 0.09 cd	LOQ	50.34 ± 2.75 a	452.24 ± 23.89 f	0.25 ± 0.04 a	***
		Wine	2.12 ± 0.72 g	8.87 ± 0.52 def	0.38 ± 0.16 c	0.06 ± 0.03 b	0.06 ± 0.03 f	LOQ	21.05 ± 0.65 f	292.88 ± 5.61 j	LOQ f	***
	Italian Riesling	Must	24.08 ± 1.23 c	19.12 ± 2.32 b	1.32 ± 0.25 a	0.20 ± 0.04 b	0.41 ± 0.19 cd	LOQ	49.87 ± 2.36 a	453.56 ± 7.85 f	0.18 ± 0.06 bcd	***
		Wine	1.41 ± 0.34 g	5.98 ± 1.28 fg	0.18 ± 0.04 cd	0.03 ± 0.02 b	0.08 ± 0.04 f	LOQ	23.86 ± 2.51 ef	299.52 ± 8.77 j	LOQ f	***
Simleul Silvaniei	Feteasca alba	Must	7.53 ± 0.06 ef	7.93 ± 0.75 ef	0.18 ± 0.05 cd	LOQ b	0.32 ± 0.05 de	LOQ	25.31 ± 3.41 ef	731.34 ± 9.84 a	0.17 ± 0.07 bcd	***
		Wine	0.25 ± 0.01 g	1.95 ± 0.06 g	LOQ d	LOQ b	0.02 ± 0.02 f	LOQ	11.59 ± 1.20 g	412.55 ± 0.61 h	LOQ f	***
	Feteasca regala	Must	8.36 ± 0.70 e	5.02 ± 0.87 fg	0.22 ± 0.03 cd	LOQ b	0.49 ± 0.04 bc	LOQ	23.58 ± 3.22 ef	711.78 ± 1.93 b	0.14 ± 0.05 cde	***
		Wine	0.12 ± 0.02 g	1.77 ± 0.07 g	0.04 ± 0.03 cd	LOQ b	0.03 ± 0.01 f	LOQ	10.38 ± 1.68 g	461.38 ± 4.37 f	LOQ f	***
	Italian Riesling	Must	5.65 ± 0.64 f	5.84 ± 0.31 fg	0.11 ± 0.03 cd	LOQ b	0.22 ± 0.03 e	LOQ	23.90 ± 3.00 ef	698.29 ± 8.59 b	0.13 ± 0.05 de	***
		Wine	0.41 ± 0.02 g	1.54 ± 0.04 g	LOQ d	LOQ b	0.02 ± 0.01 f	LOQ	13.94 ± 0.62 g	449.33 ± 6.06 f	LOQ f	***
Average		Must	19.25 ± 9.85	14.51 ± 8.41	0.61 ± 0.47	0.08 ± 0.09	0.50 ± 0.28	-	37.38 ± 11.53	589.31	0.18 ± 0.04	-
		Wine	1.17 ± 0.81	4.77 ± 2.48	0.18 ± 0.14	0.03 ± 0.02	0.05 ± 0.02	-	23.21 ± 9.77	464.15 ± 150.89	0.01 ± 0.04	-
	Sig		***	***	***	***	***	-	***	***	-	
Must	Areas		**	in	in	in	in	-	in	in	in	
	Variety		in	in	in	in	in	-	in	in	in	
	Areas x Variety		in	in	in	in	in	-	in	in	in	
Wine	Areas		***	***	***	***	***	-	***	***	***	
	Variety		in	**	*	in	in	-	in	in	***	
	Areas x Variety		in	**	in	in	in	-	in	in	***	

Table 3. *Cont.*

Areas	Variety	Sample	Cu mg/L M.P.L. 1 mg/L	Zn mg/L M.P.L. 5 mg/L	Pb mg/L M.P.L. 0.15 mg/L	Cd mg/L M.P.L. 0.01 mg/L	Ni mg/L M.P.L. -	Co µg/L M.P.L. -	As µg/L M.P.L. 0.2 mg/L	Cr µg/L M.P.L. -	Hg µg/L M.P.L. -	Sig
						Must						
	Bora et al. 2015 [23]		1.97 ± 0.78	2.70 ± 1.66	0.20 ± 0.02	LOQ	0.22 ± 0.03	-	-	-	-	-
						Wine						
	Bora et al. 2018 [24]		0.91 ± 0.04 mg/L	3268.00 ± 14.57 µg/L	125.35 ± 6.10 µg/L	0.39 ± 0.02 µg/L	682.82 ± 7.88 µg/L	7.77 ± 0.53 µg/L	14.26 ± 0.53 µg/L	620.04 ± 5.44 µg/L	0.58 ± 0.04 µg/L	-

Average value ± standard deviation ($n = 3$). Roman letters are significance of difference ($p \leq 0.05$) between the plant parts of the same variety. The difference between any two values, followed by at least one common letter, is insignificant.in = insignificant. M.P.L. – maximum permissible limit (OIV, 2005). LOQ for Pb: 0.0010 µg/L; LOQ for Cd: 0.0073 µg/L; LOQ for Co: 0.1215 µg/L. LOQ for Hg: 0.1379 µg/L.

Concentrations of Pb in must were significantly higher in Baia Sprie area than in the other two areas. All varieties cultivated in Baia Sprie and grapes of Feteasca regala from Baia Mare slightly exceeded the M.P.L. (0.5 mg/L). Concentrations are higher than those reported for Brazilian grapes juice (0.07 ± 0.00 µg/100 mL grape juice - Cabernet Sauvignon; 0.11 ± 0.00 µg/100 mL grape juice - Merlot) [33], but lower for grapes juice originated from polluted and nonpolluted regions from Serbia (1.81 ± 0.15 mg/kg) [35]. Grapevine can accumulate small amounts of Pb (27–125 mg/kg), with an average of 58.2 mg/kg in grapes [36].

The highest concentrations of Cd in must were recorded in varieties cultivated in Baia Sprie, significantly higher than in Baia Mare. Grapes samples from Simleul Silvaniei had Cd concentrations below the limit of detection. Cd is a natural component of must as it originates from the grapes. During fermentation, up to 90% of Cd accumulates in yeast, thus wine contains 0001–0002 mg/L [36]. Interestingly, must of Feteasca alba variety cultivated in Baia Mare had remarkably higher concentration of Ni compared to other varieties or the same variety from other areas (Table 3). Our values are higher than those obtained for Brazilian grapes (0.40 ± 0.01 µg/100 mL grapes juice - Cabernet Sauvignon; 0.69 ± 0.00 µg/100 g mL grapes juice - Merlot) and lower than concentrations reported for grape berries juice from Serbia (2.16 ± 0.78 mg/kg and 1.77 ± 0.14 mg/kg, respectively) [35]. The level of Co, in must, is under the detection limit in all analyzed samples. As is usually present in must as a consequence of herbicides and insecticides used for grape production, processing factors, and must storage conditions [37]. Feteasca regala and Italian Riesling varieties from Baia Mare and Baia Sprie had significantly higher concentration of As in must samples (48.30 ± 1.27 µg/L (Feteasca regala); 46.35 ± 2.60 µg/L (Italian Riesling) from Baia Mare and 50.34 ± 2.75 µg/L(Feteasca regala); 49.87 ± 2.36 µg/L (Italian Riesling) from Baia Sprie) than Feteasca alba variety from the same areas (33.06 ± 1.58 µg/L and 35.68 ± 3.29 µg/L, respectively). Concentrations of As are below the M.P.L. in all tested must samples. Highest concentrations of Cr were recorded in must samples from Simleul Silvaniei for all three varieties, while Hg was detected in comparable amounts.

2.3.2. Metal Concentration in Wine

Concentrations of Cu and Zn exceeded the M.P.L. under applicable law (1 mg/L for Cu and 5 mg/L for Zn) for varieties cultivated in Baia Sprie and Baia Mare and were below the M.P.L. in varieties from Simleul Silvaniei (Table 3). These concentrations are higher than those obtained in wine samples from different wine-producing areas of Romania: 403.92 µg/L (Cu) and 1183.32 µg/L (Zn) in Cabernet Sauvignon from Muntenia [38]; 886.31 µg/L (Cu) and 524.65 µg/L (Zn) from Muntenia, 289.52 µg/L (Cu) and 488.20 µg/L (Zn) from Dobrogea, and 642.60 µg/L (Cu) and 426.40 µg/L (Zn) from Moldova [26]. M.P.L. for Pb concentration in wine (0.15 mg/L) was exceeded in varieties from Baia Sprie and Baia Mare, the highest value being detected in Feteasca regala variety (0.38 ± 0.16 mg/L and 0.27 ± 0.02 mg/L, respectively). In other wine-producing regions, concentration of Pb was reported at 27.36 µg/L (Feteasca Neagra, Dealu-Mare) [39], 44.68 µg/L (Muntenia), 31.93 µg/L (Dobrogea), and 49.59 µg/L (Moldova) [26]. Concentrations of Cd in wine samples from Baia Sprie and Baia Mare were recorded within 0.02–0.06 mg/L, slightly above the M.P.L (0.01 mg/L); no statistically significant differences were observed between these values. In Simleul Silvaniei area, Cd concentrations were below the detection limit. Compared to our values, much lower Cd concentrations were reported for several red wine samples from Banat, Muntenia, Oltenia, and Dobrogea regions [39]. Concentration of Ni was statistically comparable in all three areas, values varying slightly between 0.02 mg/L (Simleul Silvaniei) and 0.08 mg/L (Baia Sprie). In comparison, in other Romanian wine-producing regions, similar values were reported in white wine samples but higher concentrations for red wines [26,39]. Co levels in wine samples were below the detection limit of the analytical method. Concentrations of As varied significantly amongst areas, for all three varieties, following the trend Baia Mare >Baia Sprie >Simleul Silvaniei, however, all values were below the M.P.L. imposed by law. In case of Cr, the trend was as follows: Baia Mare >Simleul Silvaniei >Baia Sprie. Concentrations of Hg were below the detection limit, except for Feteasca alba from Baia Mare (0.11 ± 0.02 µg/L).

2.4. Pearson's Correlations Between the Content of the Investigated Elements From Soil, Plant Material, Must, and Wine

The results of Pearson's correlation analysis revealed that there is a good negative correlation between metals contents in all plant parts and the distance from the "Romplumb" and "Cuprom" smelters, except for Cr and Ni in cane, Cr, Pb, and Ni in leave, Pb, As, Ni, and Co in grape, and Cd in must and wine (Table 4). Ni content in soil correlates positively with the distance. These results demonstrate that pollution resulted from metallurgical activities affect the heavy metal content of plant parts. Content of Cu, Zn, Cd, As, Pb and Hg in all plant parts decreased as the distance from the main pollution source increased, except for Ni content. Apparently, the Co smelter is not necessarily a dominant source of pollution for Pb, Co, Cr and As. These elements can be easily assimilated from soil naturally enriched with heavy metals and could come from combustion of fossil fuels in residential areas, heavy traffic, or some agricultural practices in rural zone [21,25].

Table 4. Pearson's correlation between the contents of the investigated element in plants parts and distance, between the contents of elemental in plants parts and related contents in soil, and between content in individual organs.

Metal	Distance	Metal Soil	Root	Cane	Leave	Grape	Must	Wine
Cu	–	–	–	–	–	–	–	–
Soil	−0.5148*	1.000	–	–	–	–	–	–
Root	−0.6874**	0.9942**	1.000	–	–	–	–	–
Cane	−0.6139**	0.9337**	0.8897**	1.000	–	–	–	–
Leave	−0.5106*	0.9112**	0.8616**	0.9983**	1.000	–	–	–
Grape	−0.4234*	0.9983**	0.9863**	0.9529**	0.9336**	1.000	–	–
Must	−0.6806**	0.9986**	0.9872**	0.9511**	0.9315**	0.9999**	1.000	–
Wine	−0.4786*	0.9847**	0.9603**	0.9817**	0.9817**	0.9932**	0.9925**	1.000
Zn	–	–	–	–	–	–	–	–
Soil	−0.4874*	1.000	–	–	–	–	–	–
Root	−0.6517**	0.7246**	1.000	–	–	–	–	–
Cane	−0.6542**	0.9887**	0.6133**	1.000	–	–	–	–
Leave	−0.4519*	0.9805**	0.8458**	0.9401**	1.000	–	–	–
Grape	−0.7561**	0.8136**	0.9902**	0.7175**	0.9120**	1.000	–	–
Must	−0.6325**	0.9586**	0.4985*	0.9904**	0.8841**	0.6145**	1.000	–
Wine	−0.4123*	0.9905**	0.8125**	0.9588**	0.9982**	0.8859**	0.9104**	1.000
Pb	–	–	–	–	–	–	–	–
Soil	−0.5895*	1.000	–	–	–	–	–	–
Root	−0.5587*	0.2024	1.000	–	–	–	–	–
Cane	−0.4023*	0.2490	0.9989**	1.000	–	–	–	–
Leave	−0.3655	0.4885*	0.9534**	0.9667**	1.000	–	–	–
Grape	−0.2306	0.3946	0.9797**	0.9882**	0.9945**	1.000	–	–
Must	−0.6302**	−0.4293*	0.7976**	0.7679**	0.5784*	0.6605**	1.000	–
Wine	−0.6115**	0.3369	0.9903**	0.9958**	0.9861**	0.9981**	0.7058**	1.000
Cd	–	–	–	–	–	–	–	–
Soil	−0.6003**	1.000	–	–	–	–	–	–
Root	−0.7654**	0.6589**	1.000	–	–	–	–	–
Cane	−0.5895**	0.9737**	0.8129**	1.000	–	–	–	–
Leave	−0.6012**	0.9284**	0.3321	0.8194**	1.000	–	–	–
Grape	−0.4517*	−0.3701	−0.9427**	−0.5719*	0.0017	1.000	–	–
Must	−0.3561	0.8844**	0.9338**	0.9675**	0.6476**	−0.7608**	1.000	–
Wine	−0.3328	0.6310**	−0.1679	0.4377*	0.8741**	0.4873*	0.1960	1.000

*Correlation is significant at the 0.05 level (two-tailed); **Correlation is significant at the 0.01 level (two-tailed).

Table 4. Cont.

Metal	Distance	Metal Soil	Pearson's correlation coefficients					
			Root	Cane	Leave	Grape	Must	Wine
Ni	–	–	–	–	–	–	–	–
Soil	0.5145*	1.000	–	–	–	–	–	–
Root	−0.5655*	0.6589**	1.000	–	–	–	–	–
Cane	−0.3624	0.9737**	0.8129**	1.000	–	–	–	–
Leave	−0.2784	0.9284**	0.3321	0.8194**	1.000	–	–	–
Grape	−0.3652	−0.3701	−0.9427**	−0.5719*	0.0017	1.000	–	–
Must	−0.6459**	0.8844**	0.9338**	0.9675**	0.6476**	−0.7608**	1.000	–
Wine	−0.7412**	0.6310**	−0.1679	0.4377*	0.8741**	0.4873*	0.1960	1.000
Co	–	–	–	–	–	–	–	–
Soil	−0.6874**	1.000	–	–	–	–	–	–
Root	−0.5166*	0.9767**	1.000	–	–	–	–	–
Cane	−0.5894*	−0.1765	0.0388	1.000	–	–	–	–
Leave	−0.4326*	0.5882**	0.7480**	0.6822**	1.000	–	–	–
Grape	−0.3621	−0.6345**	−0.4539*	0.8728**	0.2519	1.000	–	–
Must	0	0	0	0	0	0	1.000	–
Wine	0	0	0	0	0	0	0	1.000
As	–	–	–	–	–	–	–	–
Soil	−0.5632*	1.000	–	–	–	–	–	–
Root	−0.5132*	0.8541**	1.000	–	–	–	–	–
Cane	−0.4006*	0.4093*	−0.1249	1.000	–	–	–	–
Leave	−0.4539*	0.9815**	0.7387**	0.5766*	1.000	–	–	–
Grape	−0.3165	0.9527**	0.9718**	0.1126	0.8768**	1.000	–	–
Must	−0.6845**	0.9655**	0.6893**	0.6327**	0.9975**	0.8407**	1.000	–
Wine	−0.6632**	0.6529**	0.1638	0.9583**	0.7860**	0.3918	0.8276**	1.000
Cr	–	–	–	–	–	–	–	–
Soil	−0.5123*	1.000	–	–	–	–	–	–
Root	−0.5894*	0.9511**	1.000	–	–	–	–	–
Cane	−0.3360	0.9644**	0.9989**	1.000	–	–	–	–
Leave	−0.2135	0.8558**	0.9737**	0.9621**	1.000	–	–	–
Grape	−0.6884**	−0.9254**	0.7637**	−0.7924**	−0.5960*	1.000	–	–
Must	−0.7123**	−0.8724**	−0.6788**	−0.7122**	−0.4938*	0.9926**	1.000	–
Wine	−0.6054**	−0.0196	0.2901	0.2453	0.5004	0.3970	0.5058*	1.000
Hg	–	–	–	–	–	–	–	–
Soil	−0.7456**	1.000	–	–	–	–	–	–
Root	−0.7023**	0	1.000	–	–	–	–	–
Cane	−0.6648**	0.9912**	0	1.000	–	–	–	–
Leave	−0.7123**	0.9799**	0	0.9449**	1.000	–	–	–
Grape	−0.6948**	0	0	0	0	1.000	–	–
Must	−0.5123**	0.9527**	0	0.9042**	0.9942**	0	1.000	–
Wine	−0.7123**	−0.3812	0	−0.500*	−0.1890	0	−0.822**	1.000

*Correlation is significant at the 0.05 level (two-tailed); **Correlation is significant at the 0.01 level (two-tailed).

Significant positive correlation between metal level in plant and soil was detected in nearly all cases, while Co in grape, Pb in must, and Cr in grape and must showed significant negative correlations (Table 4). Although all elements in all samples, except for Cd, Ni, Co in grape and Cr in must, correlated positively with the metal content in roots, only the correlation of grape and root can be of interest as these organs reflect a real bioaccumulation [21].

Overall, the Pearson's correlation matrix for individual elements in soil, plant material, must, and wine showed a good positive correlation between contents of individual elements (Supplementary Table S2). Ni content in soil and Cd content in grape had negative correlation with other elements. Similar results regarding Ni behavior have been reported from Serbia [21]. The low correlation coefficients observed for Ni in soil and plant parts (except leaves) might indicate that this element comes from different sources: Ni concentration in soil is impacted predominantly by geology, and the soil is mainly the source of Ni in plants parts. Leaves of grapevine from Baia Mare and Baia Sprie have captured Ni from atmospheres as well, originating from metallurgical activities. It is a known fact that above-ground plant parts assimilate elements from both soil and atmosphere, however, leaves are likely to be the most sensitive to air pollution.

2.5. Translocation Factor (TF) and Mobility of the Element Content in the Soil-Grapevine-Wine System

TF of the metals from the soil to the aerial parts of the plant represent an essential indicator of heavy metal mobility and translocation to the edible parts of the plant. Mobility ratio (MR) in *Vitis vinifera* L. was used to determine the ratio between the metal concentration in plant parts (canes, leaves and grapes) and the concentration levels of the acid-soluble metal faction in top soil. MR >1 indicates that the plants enrich these elements (accumulator), a ratio at around 1 indicates a rather indifferent behavior of the plant towards these elements (indicator) and a ratio clearly < 1 shows that the plant exclude these elements from uptake (excluder) [40].

Mean values of TF and MR indicated effective translocation of most elements in *Vitis vinifera* L. at all three sampling sites (Tables 5 and 6). Effective translocation of Ni (Feteasca alba), Co (Feteasca alba, Feteasca regala and Italian Riesling), As (Feteasca alba, Feteasca regala and Italian Riesling), Cr (Feteasca alba, Feteasca regala and Italian Riesling) occurs from soil to grapevine roots. From roots to canes, effective translocation was recorded for Pb (Feteasca alba, Feteasca regala and Italian Riesling), Cd (Feteasca alba, Feteasca regala and Italian Riesling), Ni (Italian Riesling), Co (Feteasca regala and Italian Riesling). From canes to leaves, translocation was recorded to Cu (Feteasca alba), Pb (Feteasca alba, Feteasca regala and Italian Riesling), Ni (Feteasca alba and Feteasca regala), Co (Feteasca alba), As (Feteasca alba, Feteasca regala and Italian Riesling) and Hg (Feteasca alba, Feteasca regala and Italian Riesling), while from grapes to must, effective translocation of Cu (Feteasca alba, Feteasca regala and Italian Riesling), Zn (Feteasca alba, Feteasca regala and Italian Riesling) and Cr (Feteasca alba, Feteasca regala and Italian Riesling) was detected. For most elements, translocation coefficient between grapes-cane, must-grapes, and wine-must had values lower than 1, indicating grapevine's specific mechanisms to block the accumulation of toxic metals in grapes [41–43]. The physico-chemical and biological processes that occur in the process of transformation the must into wine generates the reducing of the heavy metals concentrations, and this is demonstrated with the lower values of the analyzed metals in wine and in must as well from the values lower than 1 of the TFs [23] based on MR values, absorption of Cu, Zn, Pb from soil to roots, canes, leaves, grapes, must, and wine of all varieties of *Vitis vinifera* L. was not considerable (MR<1). In case of Cd (canes/soils), As (roots/soil and leaves/soil), Hg (canes/soil), MR value around 1 indicates that plants had an indifferent behavior against these elements. According to literature data, *Vitis vinifera* L. can be considerate as a bioaccumulator of Pb, Cu, and Zn [14,21]. Our results also demonstrated that *Vitis vinifera* L. is not a hyperaccumulator of Cu, Zn, Pb, Cd, Ni, Co, As, Cr and Hg (absorb metals above established background concentration).

Table 5. The mean values of translocation factors in system soil-grape-wine.

Variety	Cu	Zn	Pb	Cd	Ni	Co	As	Cr	Hg	TF*** Roots/Soils
Feteasca alba	0.41	0.20	0.05	0.53	1.00	1.30	1.02	2.68	0.94	–
Feteasca regala	0.42	0.19	0.05	0.64	0.99	2.90	1.23	3.46	0.76	–
Riesling italian	0.53	0.16	0.05	0.57	0.48	2.32	1.64	2.57	0.10	–
Average	0.45 g	0.18 h	0.05 i	0.58 f	0.82 d	2.17 b	1.30 c	2.90 a	0.60 e	Cr>Co>As>Ni>Hg>Cd>Cu>Zn>Pb
STDEV*	0.07	0.02	0.00	0.06	0.30	0.81	0.31	0.48	0.44	–
RSD %**	15.25 f	12.67 g	4.39 i	9.66 h	36.23 c	37.15 b	24.01 d	16.66 e	73.01 a	Hg>Co>Ni>As>Cr>Cu>Zn>Cd>Pb
						TF Canes/Roots				
Feteasca alba	0.19	0.86	1.12	1.85	0.79	0.79	0.74	0.27	1.13	–
Feteasca regala	0.20	0.99	1.21	1.70	0.93	2.98	0.56	0.16	1.57	–
Riesling italian	0.16	1.20	1.15	1.65	1.65	3.07	0.57	0.25	1.53	–
Average	0.18 i	1.02 f	1.16 d	1.73 b	1.12 e	2.28 a	0.62 g	0.23 h	1.41 c	Co>Cd>Hg>Pb>Ni>Zn>As>Cr>Cu
STDEV	0.02	0.17	0.05	0.11	0.46	1.29	0.10	0.06	0.24	–
RSD %	12.16 g	16.49 e	4.02 i	6.19 h	41.07 b	56.77 a	16.32 f	24.99 c	17.16 d	Co>Ni>Cr>Hg>Zn>As>Cu>Cd>Pb
						TF Leaves/Canes				
Feteasca alba	1.06	0.94	1.24	0.64	3.81	3.81	2.73	0.50	2.00	–
Feteasca regala	0.96	0.79	1.05	0.74	3.23	0.10	2.46	0.72	1.76	–
Riesling italian	0.94	0.85	1.38	0.66	0.66	0.12	1.50	0.31	2.04	–
Average	0.98 f	0.86 g	1.22 e	0.68 h	2.57 a	1.35 d	2.23 b	0.51 i	1.93 c	Ni>As>Hg>Co>Pb>Cu>Zn>Cd>Cr
STDEV	0.06	0.08	0.17	0.05	1.67	2.13	0.65	0.20	0.15	–
RSD %	6.59 i	9.13 f	13.67 e	7.93 g	65.20 b	158.49 a	29.10 d	40.02 c	7.76 h	Co>Ni>Cr>As>Pb>Zn>Cd>Hg>Cu
						TF Grapes/Canes				
Feteasca alba	0.10	0.06	0.10	0.16	0.21	0.21	0.63	0.21	0.53	–
Feteasca regala	0.12	0.05	0.08	0.14	0.23	0.05	0.58	0.25	0.41	–
Riesling italian	0.11	0.05	0.08	0.15	0.15	0.07	0.38	0.24	0.65	–
Average	0.11 e	0.05 h	0.09 g	0.15 d	0.20 c	0.11 f	0.53 a	0.24 b	0.53 a	As>Hg>Cr>Ni>Cd>Cu>Co>Pb>Zn
STDEV	0.01	0.01	0.01	0.01	0.04	0.09	0.13	0.02	0.12	–
RSD %	6.34 i	13.21 f	14.13 e	7.89 h	20.70 d	81.45 a	24.81 b	9.25 g	22.39 c	Co>As>Hg>Ni>Pb>Zn>Cr>Cd>Cu

Table 5. Cont.

	Cu	Zn	Pb	Cd	Ni	Co	As	Cr	Hg	Order
TF Must/Grapes										
Feteasca alba	2.78	2.18	0.10	0.11	0.57	0.00	0.05	4.52	0.02	—
Feteasca regala	1.94	2.41	0.14	0.11	0.32	0.00	0.08	4.93	0.02	—
Riesling italian	1.84	2.24	0.15	0.11	0.11	0.00	0.09	4.16	0.02	—
Average	2.19 c	2.28 b	0.13 e	0.11 f	0.34 d	0.00 i	0.07 g	4.54 a	0.02 h	Cr>Zn>Cu>Ni>Pb>Cd>As>Hg>Co
STDEV	0.51	0.12	0.03	0.00	0.23	0.00	0.02	0.39	0.00	—
RSD %	23.48 d	5.34 h	19.65 e	3.38 i	68.35 b	173.21 a	28.83 c	8.50 g	17.75 f	Co>Ni>As>Cu>Pb>Hg>Cr>Zn>Cd
TF Wine/Must										
Feteasca alba	0.07	0.30	0.33	0.38	0.09	0.09	0.79	0.85	0.23	—
Feteasca regala	0.06	0.35	0.35	0.39	0.12	0.00	0.58	0.82	0.00	—
Riesling italian	0.06	0.35	0.22	0.22	0.22	0.00	0.60	0.89	0.00	—
Average	0.06 h	0.33 c	0.30 e	0.33 d	0.14 f	0.03 i	0.66 b	0.85 a	0.08 g	Cr>As>Zn>Cd>Pb>Ni>Hg>Cu>Co
STDEV	0.00	0.03	0.07	0.10	0.07	0.05	0.11	0.03	0.13	—
RSD %	6.16 h	9.47 g	22.67 e	29.76 d	48.64 c	173.21 a	17.52 f	3.89 i	173.21 b	Co>Hg>Ni>Cd>Pb>As>Zn>Cu>Cr

STDEV* = Standard deviation; RDS %** = Relative standard deviation; Tf*** = Translocation factors.

Table 6. The mean values of mobility ratio in system soil-grape-wine.

Variety	MR*** Roots/Soils									
	Cu	Zn	Pb	Cd	Ni	Co	As	Cr	Hg	
Feteasca alba	0.411	0.201	0.055	0.529	0.995	1.302	1.025	2.677	0.938	—
Feteasca regala	0.418	0.187	0.055	0.639	0.967	2.895	1.230	3.457	0.759	—
Riesling italian	0.534	0.156	0.051	0.566	0.477	2.317	1.637	2.571	0.105	—
Average	0.45 g	0.18 h	0.05 i	0.58 f	0.82 d	2.17 b	1.30 c	2.90 a	0.60 e	Cr>Co>As>Ni>Hg>Cd>Cu>Zn>Pb
STDEV*	0.07	0.02	0.00	0.06	0.30	0.81	0.31	0.48	0.44	—
RSD %**	15.25 f	12.67 g	4.39 i	9.66 h	36.23 c	37.15 b	24.01 d	16.66 e	73.01 a	Hg>Co>Ni>As>Cr>Cu>Zn>Cd>Pb

Table 6. *Cont.*

										Order
MR Canes/Soils										
Feteasca alba	0.076	0.174	0.062	0.981	0.782	1.023	0.758	0.725	1.062	—
Feteasca regala	0.085	0.185	0.066	1.083	0.922	8.631	0.685	0.564	1.191	—
Riesling italian	0.085	0.186	0.058	0.933	0.786	7.114	0.936	0.642	0.160	—
Average	0.082 h	0.182 g	0.062 i	0.999 b	0.830 d	5.589 a	0.793 e	0.644 f	0.805 d	Co>Cd>Ni>Hg>As>Cr>Zn>Cu>Pb
STDEV	0.005	0.007	0.004	0.077	0.080	4.026	0.129	0.080	0.562	—
RSD %	6.054 h	3.873 i	6.210 g	7.670 f	9.590 e	72.039 b	16.278 c	12.468 d	69.793 b	Hg>Co>As>Cr>Ni>Cd>Pb>Cu>Zn
MR Leaves/Soils										
Feteasca alba	0.080	0.163	0.076	0.625	2.980	3.897	2.071	0.362	2.123	—
Feteasca regala	0.081	0.145	0.069	0.802	2.978	0.896	1.685	0.407	2.099	—
Riesling italian	0.079	0.159	0.081	0.619	0.522	0.889	1.401	0.201	0.327	—
Average	0.080 h	0.156 g	0.075 i	0.682 e	2.160 a	1.894 b	1.719 c	0.323 f	1.516 d	Ni>Co>As>Hg>Cd>Cr>Zn>Cu>Pb
STDEV	0.001	0.010	0.006	0.104	1.419	1.735	0.336	0.108	1.030	—
RSD %	1.128 i	6.128 h	7.652 g	15.270 f	65.687 c	91.589 a	19.554 e	33.510 d	67.924 b	Co>Hg>Ni>Cr>As>Cd>Pb>Zn>Cu
MR Grapes/Soils										
Feteasca alba	0.008	0.011	0.006	0.159	0.165	0.215	0.480	0.153	0.568	—
Feteasca regala	0.010	0.010	0.005	0.150	0.214	0.422	0.400	0.143	0.494	—
Riesling italian	0.009	0.009	0.005	0.142	0.120	0.474	0.359	0.157	0.105	—
Average	0.009 h	0.010 g	0.005 i	0.150 f	0.166 d	0.371 c	0.413 a	0.151 e	0.389 b	As>Hg>Co>Ni>Cr>Cd>Zn>Cu>Pb
STDEV	0.001	0.001	0.001	0.008	0.047	0.137	0.062	0.007	0.249	—
RSD %	11.519 f	9.412 g	14.590 e	5.588 h	28.296 c	37.006 b	14.941 d	4.860 i	63.947 a	Hg>Co>Ni>As>Pb>Cu>Zn>Cd>Cr
MR Must/Soils										
Feteasca alba	0.022	0.023	0.001	0.018	0.094	0.000	0.024	0.693	0.009	—
Feteasca regala	0.019	0.024	0.001	0.016	0.069	0.000	0.033	0.705	0.009	—
Riesling italian	0.017	0.020	0.001	0.016	0.014	0.000	0.032	0.653	0.002	—
Average	0.020 e	0.022 d	0.001 h	0.017 f	0.059 b	0.000 i	0.030 c	0.684 a	0.007 g	Cr>Ni>As>Zn>Cu>Cd>Hg>Pb>Co
STDEV	0.003	0.002	0.000	0.001	0.041	0.000	0.005	0.027	0.004	—
RSD %	12.823 e	9.398 g	9.761 f	7.217 h	69.963 b	173.205 a	16.990 d	3.930 i	56.428 c	Co>Ni>Hg>As>Cu>Pb>Zn>Cd>Cr

Table 6. *Cont.*

MR Wine/Soils

Feteasca alba	0.001	0.007	0.000	0.007	0.008	0.000	0.019	0.586	0.002	—
Feteasca regala	0.001	0.008	0.000	0.006	0.008	0.000	0.019	0.578	0.000	—
Riesling italian	0.001	0.007	0.000	0.003	0.003	0.000	0.019	0.579	0.000	—
Average	0.001 e	0.007 c	0.000 f	0.006 d	0.006 d	0.000 f	0.019 b	0.581 a	0.001 e	Cr>As>Zn>Cd>Ni>Cu>Hg>Pb>Co
STDEV	0.000	0.001	0.000	0.002	0.003	0.000	0.000	0.004	0.001	—
RSD %	18.969 e	11.139 f	27.186 d	33.072 c	46.861 b	173.205 a	1.547 g	0.751 h	173.205 a	Co>Hg>Ni>Cd>Pb>Cu>Zn>As>Cr

STDEV* = Standard deviation; RDS %** = Relative standard deviation; MR*** = Mobility ratio.

2.6. Combining Multielement Analysis of Must and Wine for Geographical Discrimination

Elements like Mn, Cd, Li, Ba, Ca, Bi, Rb, Mg, Ag, Ni, Cr, Sr, Zn, Rb and Fe showed a high discriminatory power for geographic origin of Romanian wine, but additional new elements (Hg, Ag, As, Al, Tl, U), metal ratios (Ca/Sr and K/Rb) and $^{207}Pb/^{206}Pb$, $^{208}Pb/^{206}Pb$, $^{204}Pb/^{206}Pb$, $^{87}Sr/^{86}Sr$ isotope ratios have been investigated in order to identify new tracers for geographical traceability of Romanian wines [24,26,44].

This is the first study to assess the geographic fingerprinting of wine and must samples from a polluted area (Baia Mare and Baia Sprie). The analyzed wine samples showed high concentration of elements, but not exceeding the maximum levels recommended by International Organisation of Vine and Wine (OIV 2016), except for Cu, Zn, Pb and Cd in Baia Mare and Baia Sprie. In Simleul Silvaniei, the high concentration of some elements is mostly derived from agricultural practices, fertilizers, and technological winemaking processes. Multivariate chemometric method was applied for the differentiation of must and wine intro groups based on their geographic origin. Linear discriminant analysis (LDA) was used to identify significant tracers for classification to the geographical discrimination of the wine samples.

Based on the elemental contents, cross-validation technique provided an 88.09% and 84.87% percentage of predicted membership according to the must and wine geographic origin, respectively (Supplementary Figures S1 and S3). The linear correction revealed acceptable scores for the two defined discriminant factors (F1 = 73.09%, F2 = 15.01% for must and F1 = 62.36%, F2 = 22.50% for wine). F1 mainly separates Baia Mare and Baia Sprie areas from Simleul Silvaniei and F2 separates Simleul Silvaniei from Baia Mare and Baia Sprie (Supplementary Figure S2). Among the investigated parameters, Cr, Hg, As, Cu, Zn, Pb, Ni and Cd was identified as the most significant for geographic differentiation of the must and wine from Baia Mare, Baia Sprie, and Simleul Silvaniei areas. The technique of cross-validation was applied during the set validation and the proposed model appears to be a promising chemometric approach for precise classification of wines according to their geographical origin. Thus, in both cases, the geographical regions were correctly classified with percentage between 52% and 71%.

2.7. Cluster Analysis

The hierarchical dendrogram for polluted sites based on elements content in sol material (Supplementary Figure S5) showed two primary clusters of the contaminated locations. The first cluster is formed of sites located in Simleul Silvaniei area, while the second one is formed of sites from Baia Mare and Baia Sprie. In terms of measure interval, the difference between the two primary clusters was significant, which suggests higher soil pollution in Baia Mare and Baia Sprie compared to Simleul Silvaniei. Both primary clusters were further divided into several new subclusters. However, the differentiation between the areas from Baia Mare and Baia Sprie was more significant than Simleul Silvaniei area. The position of an isolated subcluster which belongs to the Baia Mare area suggested that this area is the most polluted one. The dendrogram of elements in vineyard soil (Supplementary Figure S6) showed two main clusters (one isolated for As and other for the rest of elements) and numerous subclusters. The difference between primary clusters was significant, which confirmed the previous conclusion that the source of As content in soils is of geological origin, whereas the concentrations of other metals in soil are also influenced by atmospheric pollution. This was particularly obvious in the case of Cu, Pb, Zn and Cd. Similar conclusions can be formulated from analysis of the dendrogram based on element contents in grapevine roots (Supplementary Figure S7), that indicated one cluster for Ni and another cluster for the rest of elements, as well as numerous different subclusters. The dendrogram of elements in grapevine canes, leaves and grapes (Supplementary Figures S8–S10) showed two main cluster: one isolated for Hg (canes dendrogram), As (leaves dendrogram), and Cd (grapes dendrogram) and another for the rest of elements; and several different subclusters. These results also demonstrated the two possible sources of the investigated elements in these organs: soil or atmosphere. The hierarchical dendrogram for must and wine based on elements content

(Supplementary Figures S11 and S12) showed two primary clusters. For must, first cluster is formed by Zn, Cu, Hg, Cd, Pb, Co and As and the second cluster is formed by Cr and Ni. For wine, first cluster is formed by Pb, Zn, Cd, Cu, Ni and the second cluster is formed by Hg, Co, Cr, As. The hierarchical dendrogram for the elements in the upper organs of grapevine (Supplementary Figure S13) also showed two main clusters: one cluster formed by Co, Ni, Hg, Cd, (grapes), Hg (canes), As (leave) and other for the rest of elements, in canes, leaves and grapes, as well as numerous different subclusters which demonstrated well a fine structure with two possible sources for the investigated elements: soil or atmosphere. The grouping of the elements confirmed that the Co, Ni, Hg, Cd, As concentrations of soil are the main source of Co, Ni, Hg, Cd, As content in the upper organs and the influence of atmospheric pollution is the highest for the group consisting of: Zn grape, Cr cane, Co leave, Cr, grape, Cr leave, As cane, that are placed furthest from the primary cluster. The combination of methods used in this study for data analysis, such as the calculation of TFs, MRs, Pearson's correlation study, and hierarchical cluster analysis, provided a very valuable information that made feasible a multi-aspect construction of the grapevine study and can be recommended for any similar investigation.

3. Materials and Methods

3.1. Description of the Sampling Area

The present study was conducted in Baia Mare and Baia Sprie area, one of the important mining districts in Romania. The main mining activities previously developed in the area considered of nonferrous sulfidic ore extraction and processing, aiming to obtain concentrated of Pb, Cu, Zn and precious metals. After 2006, the metallurgical industry from Baia Mare and Baia Sprie has considerably diminished its activity by closing or reducing its production capacity.

Baia Mare depression is a contact depression the interposes between the Someșana Plain and the Carpathian Mountains as a lower morphological unit, from the surrounding areas, presenting a waved surface, characterized by a convergent system of valleys and interfluves. It was formed due to the tertiary tectonic movement that took to the fragmentation and sinking of the crystalline in the Northwest part of Transylvania, as well as due to the volcanic chain of the Gutin-Oaș Mountains. The metropolitan area of Baia Mare is in the NW of Romania, in a hilly region, at an altitude of 220 m above sea level, covering an area of 1250 km^2 and having a population of more than 200.000 residents.

The Șimleul Silvaniei vineyard is located in the northwest of Romania and is delimited by the Apuseni Carpathians on the south, the Someșan Plateau on the east and the Someșan Plain on the northwest, which is known geographically under the name of Silvaniei Hills. The altitude of this depression decreases from 500 m, in the accumulation area under the mountain, at 350–300 m, located in the wide part between the Măgura Șimleului and the Plopiș Mountains. Because of its position is among the northernmost vineyard in Romania. The climate of Baia Mare, Baia Sprie and Șimleul Silvaniei area falls in both moderate continental and the mountain climate categories [45].

3.2. Description of the soil types

According to the Romanian Soil Taxonomic Classification [46] in the investigated areas there were found: eutricambosol, typical luvosol, stagnic luvosol, gleyic luvosol, and aluviosols. Vegetation characteristic of eutricambosol soils was represented by forests partly replaced by pastures and meadows. Eutricambosols are moderate acidic with a slight difference on soil profile. Humus content is relatively high in the organic horizon (2.76–4.44%) [46]. Luvisols were represented by typical stagnic and gleyic luvosol types. They appear on a small area near metallurgical plant and are prevalent in the southern extension of the investigated areas. These soils are developed on the low plains and poorly drained terrains. Typical luvosol was present on large areas, being covered by orchards and grasslands. The Ao horizon has grey colour. The colors of the Bt horizon vary from red to brown. Soil profile was as follows: Ao-Bt-C. Stagnic and gleyic luvosol types were poor in nutrients and humus and had low natural fertility being covered by natural grasslands. Soil profile was as follows:

Ao-Ea(El)-E/B-Bt-C. The Ao horizon was 15 cm thick, the brown-grey color indicating a low content of humus. The structure was granular; the texture ranging from clay loamy to clay. Aluviosols were presented only in the western proximity of metallurgical plant and were consisted of an Ao horizon of 40 cm, which on top of C horizon of alluvial deposits [46,47].

3.3. Sample Collection and Processing

Soil, cane, and leave samples of grapevine were collected from Baia Mare, Baia Sprie and surrounding areas (Simleul Silvaniei) (Figure 1) during the vegetation period in May 2012. Soil samples were collected at the depth of 0–20, 20–40, 40–60 and 60–80 cm at the vineyard. Grapes of Feteasca alba, Feteasca regala, and Italian Riesling varieties were sampled one week before harvesting in August 2012. Roots (diameters <2.5 mm and >2.5 mm), canes (50–70 cane pieces of 25 cm), and leaves (50–70 fully-developed leaves from the middle part of the one-year old cane) were also collected. After removing damaged plant materials, samples were placed in sealed plastic bags and were immediately transported to the laboratory. Plant materials and soil samples were carefully processed to avoid chemical and physical interactions and analyzed by Inductively Coupled Plasma Mass Spectrometry (ICP-MS) Waltham, Massachusetts, SUA (see the Supplementary Materials).

Figure 1. Map of the Mining and Smelting Complex Baia Mare (Northwest Romania) with the sampling points.

3.4. Soil Sample Preparation

The soil samples (100 samples) were dried, homogenized and then passed through a 20-mesh sieve to obtain very fine particles. The method for microwave digestion using a Milestone START D Microwave Digestion System (Sorisole, Italy) was optimized in a previous work [22]: 0.25 g soil, 9 mL 65% HNO_3, 3 mL concentrate HF and 2 mL concentrated HCl were placed in a clean Teflon digestion vessel. The vessel was closed tightly and placed in the microwave. The digestion was carried out with the program described in Supplementary Table S3.

3.5. Plant Material Samples (Roots, Canes and Leaves) Preparation

The plant material samples (75 samples of roots, 113 samples of canes and 140 samples of leaves) were thoroughly washed with tap water followed by ultra-pure water using Milli-Q Integral ultrapure water-Type 1 (Darmstadt, Germany), after washing was oven-dried at 80 °C to constant weight using a FD 53 Binder (Darmstadt, Germany). The dried samples were ground using a Retsch 110 automatic mill (Darmstadt, Germany), passed through a 2 mm sieve to obtain very fine particles. The method for microwave digestion using a Milestone START D Microwave Digestion System (Sorisole, Italy) was optimized in a previous work [23]: 1 g sample of plant material, 7 mL 65% HNO_3 and 2 mL H_2O_2 were placed in a clean Teflon digestion vessel. The vessel was closed tightly and placed in the microwave. The digestion was carried out with the program described in Supplementary Table S3.

3.6. Grape Juice Sample Preparation

Grape samples (100–110 kg/cultivar) were collected from each cultivar from 70 vines. The grapes placed in the top, middle and lower third of each vine and grapes were exposed to sun and shade [22]. In this way can achieve better homogenization of sample grapes. Feteasca regala (three samples), Feteasca alba (three samples), Italian Riesling (three samples) grape juices (must) were cold pressed manually. Before the analysis, each juice samples (50 mL) were diluted in different proportions using ultrapure water. All samples were taken in triplicates from the defined experimental plot of which had a size of 5 ha.

3.7. Microvinification Process

The samples of grapes were destemmed and crushed, then transferred to a microfermentor (50 L) cylindrical glass container, covered with aluminium foil to limit the effect of the light over the must) equipped with a fermentation airlock. Fermentation took place at 22–24 °C and humidity 55–60%. Afterwards wine was clarified by means of bentonite (40 g/L 1:10 dilution) and combined with SO_2 up to 100 g/L. Then wines were allowed to cool for thirty days at −5 °C for cold stabilization [23]. Then wine samples were stored in glass bottles at 5–6 °C until the analyses. Average data from three vinifications per cultivar are reported [23].

3.8. Wine Sample

The wine samples were taken from freshly opened bottles and prepared by a specific organic matter digestion. 2.5 mL of wine were weighed inside Teflon digestion vessels and 2.5 mL concentrated HNO_3 added. Teflon digestion vessels were previously cleaned in nitric solution to avoid cross-contamination. The vessels already capped were placed in a microwave oven followed by the application of the program described in Supplementary Table S3, optimized in a previous work [23]. After cooling to ambient temperature, the microwave oven was opened and the content was quantitatively transferred into a 50 mL volumetric flask and brought to the volume with ultra-pure water. All the elements were measured from these extraction solutions by ICP-MS (Waltham, Massachusetts, SUA).

3.9. Inductively Coupled Plasma Mass Spectrometer (ICP-MS) Analysis

Analytical measurements were performed using an inductively coupled plasma mass spectrometer (iCAP Q ICP-MS Thermo Fisher Scientific, Waltham, Massachusetts, SUA) equipped with an ASX-520 autosampler, a micro-concentric nebulizer, nickel cones and peristaltic sample delivery pump, running a quantitative analysis mode. Each sample was analyzed in duplicate and each analysis consisted of seven replicates. The gaseous argon and helium used to form the plasma in the ICP-MS was of purity 6.0 (Messer – Gases for Life, Austria). The heavy metals were measured by using a multi-element analysis after appropriate dilution using an external and standard calibration. The calibration was performed using XXICertiPUR multielement standard, and from individual standard solution of Hg. The working standards and the control samples were prepared daily from the intermediate standards that were prepared from the stock solution. The intermediate solutions stored in polyethylene bottles and glassware were cleaned by soaking in 10% *v/v* HNO_3 for 24 h and rinsing at least ten rimes with ultrapure water (Milli-Q Integral ultrapure water-Type 1). The accuracy of the methods was evaluated by replicate analyses of fortified samples (10 μL–10 mL concentrations) and the obtained values ranged between 0.8–13.1%, depending on the element. The global recovery for each element was estimated and the obtained values were between 84.6–100.9%.

For quality control purpose, blanks and triplicates samples ($n = 3$) we analyzed during the procedure. The variation coefficient was under 5% and detection limits (ppb) were determined by the calibration curve method. Limit of detection (LoD) and Limit of quantification (LoQ) limits were calculated according to the next mathematical formulas: LoD = 3×SD/s and LoQ = 10×SD/s (SD = estimation of the standard deviation of the regression line; s = slope of the calibration curve) (Supplementary Table S4). The recovery assays for the must and wine sample of 5 µL concentration, for three replicates of this level of concentration ($n = 3$) gave the average recovery R % between 87.32% and 100.26%. The recovery for the soil and plant material samples of 5 µL concentration, for three replicates of this level of concentration ($n = 3$) gave the average recovery R % between 83.41% and 109.02%. Optimum instrumental conditions for ICP-MS measurement are summarized in Supplementary Table S3. The calibration standards were prepared from the multielement standard solution, ICP Multi Element Standard Solution XXI CertiPUR, in five concentration ranges 2.5, 5, 10, 25 and 50 µL.

3.10. The Determination of pH, Electrical Conductivity (EC) and Organic Matter (OM)

The pH and EC of soil samples (soil/distilled water = 1:2.5) were measured using pH meter Jenway, 3510, Keison (Chelmsford, UK) and an Electrical Conductivity (EC) meter Jenway, 3510, Keison (Chelmsford, UK), respectively. The organic matter (OM) was determined by loss-on-ignition method at 550 °C [21].

3.11. Reagents and Solutions

High purity ICP Multi-element Standard Solution XXI CertiPUR obtained from Merck (Darmstadt, Germany) was used for the calibration curve in the quantitative analysis. HNO_3, concentrated HF and HCl (reagent grade from Merck, Darmstadt, Germany) and ultra-pure water (maximum resistivity of 18.2 MΩ × cm^{-1}, Milli-Q Integral ultrapure water-Type 1) were used for sample preparation.

3.12. Statistical Analysis

Average and standard deviation were calculated, and data were interpreted with the analysis of variance (ANOVA) and the average separation was performed with the Duncan test at $p \leq 0.005$. Pearson's correlation coefficient was calculated using SPSS Version 24 (SPSS Inc., Chicago, IL, USA), Excel 2016 (Microsoft, New York, NY, USA) and Addinsoft version 15.5.03.3707 (Microsoft, New York, NY, USA). Value higher than 0.5 indicate a strong correlation between analyzed varieties, a positive correlation between two parameters shows that both parameters increased, and a negative correlation indicates that a parameter increased while the second one decreased and vice-versa. Linear discriminant analysis (LDA) was performed to separate the wines by region and to identify the markers with a significant discrimination value (variables with Wilk's lambda near zero, p values <0.005 and higher F coefficients), using Microsoft Excel 2016 and XLSTAT Addinsoft version 15.5.03.3707. By cross-validation, we established the optimal number of parameters required to obtain a robust model.

Trace metal TF in grapevine was determined by the equation (TF$_{r-s}$ = C$_{roots}$/C$_{soils}$; TF$_{c-r}$ = C$_{canes}$/C$_{roots}$; TF$_{l-c}$ = C$_{leaves}$/C$_{canes}$; TF$_{m-c}$ = C$_{must}$/C$_{canes}$; TF$_{w-m}$ = C$_{wine}$/C$_{must}$ as the ratio between roots-soil; canes-roots; leaves-canes; must-canes, and wine-must. TF > 1 indicates that grapevine translocates metals effectively from soil to plants parts [43]. The MR between the metal concentration in plant parts (C$_{plant}$, mg/kg) and concentration in the top-soil (C$_{soil-m}$, mg/kg) was determined according to the equation MR = C$_{plant}$/C$_{soil-m}$. MR > 1 indicates effective metal translocation from soil to plants parts.

4. Conclusions

All organs and products of *Vitis vinifera* L., except for grapes, must, and wine, provide numerous pieces of reliable information for efficient biomonitoring. Obtained data showed a very low environmental quality of the ecosystem in Baia Mare, Baia Sprie, and their surrounding areas. Furthermore, the content of most elements in plant parts is affected by airborne pollution which comes from nearby metallurgical activities, i.e., from the Cu smelter, whereas geology contributes predominately to the Ni content. Also, these results suggest that the Cu smelter is not necessarily a dominant source of pollution by As and Hg.

The most abundant elements in all plants, soil samples, must, and wine from Baia Mare and Baia Sprie areas were Cu and Zn, except for grape samples. Apparently, the investigated grapevine cultivar poses some specific means for a strong protection of grapes from high concentrations of heavy metals, while tolerates considerable amounts of heavy metals (Cu, Zn, Hg, As) in other tissues, especially in root tissue. This means that the *Vitis vinifera* L. cultivated in Baia Mare and Baia Sprie areas may have developed a wide range of cellular mechanisms that are highly effective in heavy metal detoxification and tolerance to heavy-metal-induced stress, including different tactics of restriction of metal uptake from soil as well as the retention of assimilated metals in the root tissue. Except of sporadic incidences, there were no visible symptoms of phytotoxic effects of metals, even though many of the grapevines were growing in highly polluted soils. Planting of the *Vitis vinifera* L. can be recommended in all kinds of soils that are severely polluted with heavy metals because it is a suitable candidate for phytostabilization. The plants of this climber species may also be useful as a vegetation protection barrier from considerable atmospheric pollution. At the same time, berries are safe for consumption to a large degree, which is a great advantage of this species.

Supplementary Materials: The following are available online at http://www.mdpi.com/1420-3049/25/3/750/s1, The physical properties of the soils samples; Table S1. The physical properties of the soils samples (Mean ± standard deviation) ($n = 3$); Table S2. Pearson's correlation matrix for investigated elemental in soil, plant material, must and wine; Figure S1. Correlation between analyzed parameters and the factors in discriminant analysis of must geographic origin; Figure S2. Differentiation of must according to geographic origin based on elements content; Figure S3. Correlation between analyzed parameters and the factors in discriminant analysis of wine geographic origin; Figure S4. Differentiation of wine according to geographic origin based on elements content; Figure S5. Hierarchical dendrogram for polluted sites based on element contents in soils; Figure S6. Hierarchical dendrogram for elements in vineyard soil; Figure S7. Hierarchical dendrogram for elements in grapevine roots; Figure S8. Hierarchical dendrogram for elements in grapevine canes; Figure S9. Hierarchical dendrogram for elements in grapevine leaves; Figure S10. Hierarchical dendrogram for elements in grapevine grapes; Figure S11. Hierarchical dendrogram for elements in grapevine must; Figure S12. Hierarchical dendrogram for elements in grapevine wine; Figure S13. Hierarchical dendrogram for elements in grapevine upper organs; Table S3. The program of the microwave oven Milestone START D Microwave Digestion System; Table S4. LoD, LoQ, BEC and r2 of the calibration for each element; Table S5. Instrumental (a) and data acquisition (b) parameters of ICP-MS.

Author Contributions: F.D.B. and C.I.B. conceived and designed the experiments; F.D.B. performed the sample collection and processing, the determination of pH, electrical conductivity and organic matter and wrote the first draft of the manuscript. C.I.B. and R.C. contributed to statistical analysis and manuscript revision. A.B. contributes to data analysis and revised the manuscript. All authors have read and agreed to the published version of the manuscript.

Funding: The publication was supported by funds from the Ministry of Research and Innovation through Program 1 - Development of the National Research and Development System, Subprogram 1.2 - Institutional Performance - Projects for Financing the Excellence in CDI, Contract no. 37PFE/06.11.2018.

Conflicts of Interest: The authors declare no conflict of interest. The funders had no role in the design of the study; in the collection, analyses, or interpretation of data; in the writing of the manuscript, or in the decision to publish the results.

References

1. Krüger, C.; Carius, A. *Environmental Policy and Law in Romania. Towards EU Accession*; Books on Demand: Berlin, Germany, 2001.
2. Muntean, E.; Muntean, N.; Duda, D. Heavy metal contamination of soil in Copșa Mică. *ProEnvironment* **2013**, *6*, 469–473.
3. Paulette, L.; Man, T.; Weindorf, D.C.; Person, T. Rapid assessment of soil and contaminant variability via portable x-ray fluorescence spectroscopy: Copșa Mică, Romania. *Geoderma* **2015**, *243–244*, 130–140. [CrossRef]
4. Maric, M.; Antonijevic, M.; Alagic, S. The investigation of the possibility for using some wild and cultivated plants as hyperaccumulators of heavy metals from contaminated soil. *Environ. Sci. Pollut. Res. Int.* **2013**, *20*, 1181–1188. [CrossRef] [PubMed]

5. Zeiner, M.; Kuhar, A.; Cindrić, I.J. Geographic differences in element accumulation in needles of Aleppo Pines (*Pinus halepensis* Mill.) grown in Mediterranean region. *Molecules* **2019**, *24*, 1877. [CrossRef]
6. Kadukova, J.; Manousaki, E.; Kalogerakis, N. Pb and Cd accumulation and phyto-excretion by salt cedar (Tamarix smyrnensis Bunge). *Int. J. Phytoremed.* **2008**, *10*, 31–46. [CrossRef]
7. Vamerali, T.; Bandiera, M.; Mosca, G. Field crops for phytoremediation of metal-contaminated land. A review. *Environ. Chem. Lett.* **2010**, *8*, 1–17. [CrossRef]
8. Palmer, C.M.; Guerinot, M.L. Facing the challenges of Cu, Fe and Zn homeostasis in plants. *Nat. Chem. Biol.* **2009**, *5*, 333–340. [CrossRef]
9. Kopittke, P.M.; Menzies, N.W.; Wang, P.; McKenna, B.A.; Wehr, J.B.; Lombi, E.; Kinraide, T.B.; Blamey, F.P. The rhizotoxicity of metal cations is related to their strength of binding to hard ligands. *Environ. Toxicol. Chem.* **2014**, *33*, 268–277. [CrossRef]
10. Kirkham, M.B. Cadmium in plants on polluted soils: Effects of soil factors, hyperaccumulation, and amendments. *Geoderma* **2006**, *137*, 19–32. [CrossRef]
11. Simon, E.; Braun, M.; Vidic, A.; Bogyó, D.; Fábián, I.; Tóthmérész, B. Air pollution assessment based on elemental concentration of leaves tissue and foliage dust along an urbanization gradient in Vienna. *Environ. Pollut.* **2011**, *159*, 1229–1233. [CrossRef]
12. Amoros, J.; Navarro, F.J.; Pérez De, C.; Pérez-de-los-Reyes, C.; Campos, J.A.; Bravo, S.; Ballesta, J.; Moreno, R. Geochemical Influence of Soil on Leaf and Grape (*Vitis vinifera* L. "Cencibel") Composition in La Mancha Region (Spain); *Vitis* **2012**, *51*, 111–118.
13. Albulescu, M.; Turuga, L.; Masu, S.; Uruioc, S.; Kiraly, L.S. Study regarding the heavy metal content (lead, chromium, cadmium) in soil and Vitis Vinifera in vineyards from the Caraș-Severin County. *Ann. West Univ. Timișoara Ser. Chem.* **2009**, *18*, 4552.
14. Vystavna, Y.; Rushenko, L.; Diadin, D.; Klymenko, O.; Klymenko, M. Trace metals in wine and vineyard environment in southern Ukraine. *Food Chem.* **2014**, *146*, 339–344. [CrossRef]
15. Damian, F.; Damian, G.; Radu, L.; Macovei, G.; Iepure, G.; Năprădean, I.; Chira, R.; Kollar, L.; Rață, L.; Dorina, C.; et al. Soils from the Baia Mare Zone and the Heavy Metals Pollution. *Carpath. J. Earth Environ. Sci.* **2008**, *3*, 85–98.
16. Levei, E.A.; Miclean, M.; Senila, M.; Cadar, O.; Roman, C.; Micle, V. Assessment of Pb, Cd, Cu and Zn availability for plants in Baia Mare mining region. *J. Plant. Dev.* **2010**, *17*, 139–144.
17. Senila, M.; Cristina, M.; Michnea, A.; Gabriela, O.; Roman, C.; Stela, J.; Butean, C.; Barz, C. Arsenic and Antimony Content in Soil and Plants from Baia Mare Area, Romania. *Am. J. Environ. Sci.* **2010**, *6*, 33–40.
18. Mihali, C.; Oprea, G.; Michnea, A.; Jelea, S.G.; Jelea, M.; Man, C.; Șenila, M.; Grigor, L. Assessment of heavy content and pollution level in sol and plants in Baia Mare area, NW Romania. *Carpath. J. Earth Environ. Sci.* **2013**, *8*, 143–152.
19. Donici, A.; Bunea, C.I.; Călugăr, A.; Harsan, E.; Racz, I.; Bora, F.D. Assessment of Heavy Metals Concentration in Soil and Plants from Baia Mare Area, NW Romania. *Bull. UASVM Horticult.* **2018**, 75. [CrossRef]
20. Huzum, R.; Iancu, G.O.; Buzgar, N. Geochemical distribution of selected trace elements in vineyard soils from the Huși area, Romania. *Carpath. J. Earth Environ. Sci.* **2012**, *7*, 61–70.
21. Alagić, S.Č.; Tošić, S.B.; Dimitrijević, M.D.; Antonijević, M.M.; Nujkić, M.M. Assessment of the quality of polluted areas based on the content of heavy metals in different organs of the grapevine (*Vitis vinifera*) cv Tamjanika. *Environ. Sci. Pollut. Res.* **2015**, *22*, 7155–7175. [CrossRef]
22. Bravo, S.; Amorós, J.A.; Pérez-de-los-Reyes, C.; García, F.J.; Moreno, M.M.; Sánchez-Ormeño, M.; Higueras, P. Influence of the soil pH in the uptake and bioaccumulation of heavy metals (Fe, Zn, Cu, Pb and Mn) and other elements (Ca, K, Al, Sr and Ba) in vine leaves, Castilla-La Mancha (Spain). *J. Geochem. Explor.* **2017**, *174*, 79–83. [CrossRef]
23. Bora, F.D.; Bunea, C.I.; Rusu, T.; Pop, N. Vertical distribution and analysis of micro-, macroelements and heavy metals in the system soil-grapevine-wine in vineyard from North-West Romania. *Chem. Cent. J.* **2015**, *9*, 19. [CrossRef] [PubMed]
24. Bora, F.D.; Donici, A.; Rusu, T.; Bunea, A.; Popescu, D.; Bunea, C.I. Elemental Profile and 207Pb/206Pb, 208Pb/206Pb, 204Pb/206Pb, 87Sr/86Sr Isotope Ratio as Fingerprints for Geographical Traceability of Romanian Wines. *Not. Bot. Horti Agrobot. Cluj-Napoca* **2018**, *46*. [CrossRef]
25. Kabata-Pendias, A. *Trace Elements in Soil and Plants*; CRC: Boca Raton, FL, USA; Washington, DC, USA, 2010.

26. Geană, E.I.; Iordache, A.; Ionete, R.; Marinescu, A.; Ranca, A.; Culea, M. Geographical origin identification of Romanian wines by ICP-MS elemental analysis. *Food Chem.* **2013**, *138*, 1125–1134. [CrossRef]
27. Lacatusu, R. Contributions regarding heavy metals flow within soil-plant-animal system in polluted areas. *Acta Met.* **2014**, *11*, 73–88.
28. Ungureanu, T.; Iancu, G.O.; Pintilei, M.; Chicoș, M.M. Spatial distribution and geochemistry of heavy metals in soils: A case study from the NE area of Vaslui county, Romania. *J. Geochem. Explor.* **2017**, *176*, 20–32. [CrossRef]
29. Chaignon, V.; Sanchez-Neira, I.; Herrmann, P.; Jaillard, B.; Hinsinger, P. Copper bioavailability and extractability as related to chemical properties of contaminated soils from a vine-growing area. *Environ. Pollut.* **2003**, *123*, 229–238. [CrossRef]
30. Toselli, M.; Baldi, E.; Marcolini, G.; Malaguti, D.; Quartieri, M.; Sorrenti, G.; Marangoni, B. Response of potted grapevines to increasing soil copper concentration. *Aust. J. Grape Wine Res.* **2009**, *15*, 85–92. [CrossRef]
31. Alloway, B.J. *Heavy Metals in Soils: Trace Metals and Metalloids in Soils and their Bioavailability*; Springer: Dordrecht, The Netherlands, 2013; Volume 22.
32. Juang, K.W.; Lee, Y.I.; Lai, H.Y.; Wang, C.H.; Chen, B.C. Copper accumulation, translocation, and toxic effects in grapevine cuttings. *Environ. Sci. Pollut. Res.* **2012**, *19*, 1315–1322. [CrossRef]
33. Panceri, C.P.; Gomes, T.M.; De Gois, J.S.; Borges, D.L.G.; Bordignon-Luiz, M.T. Effect of dehydration process on mineral content, phenolic compounds and antioxidant activity of Cabernet Sauvignon and Merlot grapes. *Food Res. Int.* **2013**, *54*, 1343–1350. [CrossRef]
34. Stefanowicz, A.M.; Stanek, M.; Woch, M.W.; Kapusta, P. The accumulation of elements in plants growing spontaneously on small heaps left by the historical Zn-Pb ore mining. *Environ. Sci. Pollut. Res.* **2016**, *23*, 6524–6534. [CrossRef] [PubMed]
35. Todic, S.; Beslic, Z.; Lakic, N.; Tesic, D. Lead, Mercury, and Nickel in Grapevine, *Vitis vinifera* L., in Polluted and Nonpolluted Regions. *Bull. Environ. Contam. Toxicol.* **2006**, *77*, 665–670. [CrossRef] [PubMed]
36. Țârdea, C. *Chimia și Analiza Vinului [Chemistry and Wine Analysis]*; Ion Ionescu de la Brad: Iași, Romania, 2007.
37. Ibanez, J.G.; Carreon-Alvarez, A.; Barcena-Soto, M.; Casillas, N. Metals in alcoholic beverages: A review of sources, effects, concentrations, removal, speciation, and analysis. *J. Food Compos. Anal.* **2008**, *21*, 672–683. [CrossRef]
38. Bora, F.D.; Donici, A.; Moldovan, M.P. Measurements of trace elements in must and wine using FASS technique. *AAB Bioflux* **2015**, *7*, 157–165.
39. Avram, V.; Magdas, D.A.; Voica, C.; Cristea, G.; Cimpoiu, C.; Hosu, A.; Marutoiu, C. Isotopic Oxygen Ratios and Trace Metal Determination in Some Romanian Commercial Wines. *Anal. Lett.* **2014**, *47*, 641–653. [CrossRef]
40. Baker, A.J.M. Accumulators and excluders-strategies in the response of plants to heavy metals. *J. Plant Nutr.* **1981**, *3*, 643–654. [CrossRef]
41. Chojnacka, K.; Chojnacki, A.; Gorecka, H.; Gorecki, H. Bioavailability of heavy metals from polluted soils to plants. *Sci. Total Environ.* **2005**, *337*, 175–182. [CrossRef]
42. Mingorance, M.D.; Valdés, B.; Oliva, S.R. Strategies of heavy metal uptake by plants growing under industrial emissions. *Environ. Int.* **2007**, *33*, 514–520. [CrossRef]
43. Serbula, S.M.; Miljkovic, D.D.; Kovacevic, R.M.; Ilic, A.A. Assessment of airborne heavy metal pollution using plant parts and topsoil. *Ecotox. Environ. Safe.* **2012**, *76*, 209–214. [CrossRef]
44. Geană, E.-I.; Sandru, C.; Stanciu, V.; Ionete, R.E. Elemental Profile and 87Sr/86Sr Isotope Ratio as Fingerprints for Geographical Traceability of Wines: An Approach on Romanian Wines. *Food Anal. Methods* **2017**, *10*, 63–73. [CrossRef]
45. Cotea, V.C.; Andreescu, F. *Romania, Wine Country*; AdLibri: Bucharest, Romania, 2009.
46. Damian, G.; Damian, F.; Năsui, D.; Pop, C.; Cornel, P. The Soils Quality from the Southern-Eastern Part of Baia Mare Zone Affected by Metallurgical Industry. *Carpath. J. Earth Env.* **2010**, *5*, 139–147.
47. Florea, N.; Munteanu, I. *The Romanian Soil Taxonomy System*; Editura Estfalia Press: Bucharest, Romania, 2003. (In Romanian)

© 2020 by the authors. Licensee MDPI, Basel, Switzerland. This article is an open access article distributed under the terms and conditions of the Creative Commons Attribution (CC BY) license (http://creativecommons.org/licenses/by/4.0/).

Article

Analysis of Pollution in High Voltage Insulators via Laser-Induced Breakdown Spectroscopy

Xinwei Wang [1], Shan Lu [1], Tianzheng Wang [1], Xinran Qin [2], Xilin Wang [2,*] and Zhidong Jia [2]

1. Shanxi Electric Power Research Institute, Taiyuan 030000, China; wxw7912@163.com (X.W.); ls8760033@163.com (S.L.); wtz2000@163.com (T.W.)
2. Engineering Laboratory of Power Equipment Reliability in Complicated Coastal Environments, Tsinghua Shenzhen International Graduate School, Shenzhen 518055, China; txr19@mails.tsinghua.edu.cn (X.Q.); jiazd@sz.tsinghua.edu.cn (Z.J.)
* Correspondence: wang.xilin@sz.tsinghua.edu.cn

Academic Editors: Clinio Locatelli, Marcello Locatelli and Dora Melucci
Received: 13 December 2019; Accepted: 10 February 2020; Published: 13 February 2020

Abstract: Surface pollution deposition in a high voltage surface can reduce the surface flashover voltage, which is considered to be a serious accident in the transmission of electric power for the high conductivity of pollution in wet weather, such as rain or fog. Accordingly, a rapid and accurate online pollution detection method is of great importance for monitoring the safe status of transmission lines. Usually, to detect the equivalent salt deposit density (ESDD) and non-soluble deposit density (NSDD), the pollution should be collected when power cut off and bring back to lab, time-consuming, low accuracy and unable to meet the online detection. Laser-induced breakdown spectroscopy (LIBS) shows the highest potential for achieving online pollution detection, but its application in high voltage electrical engineering has only just begun to be examined. In this study, a LIBS method for quantitatively detecting the compositions of pollutions on the insulators was investigated, and the spectral characteristics of a natural pollution sample were examined. The energy spectra and LIBS analysis results were compared. LIBS was shown to detect pollution elements that were not detected by conventional energy spectroscopy and had an improved capacity to determine pollution composition. Furthermore, the effects of parameters, such as laser energy intensity and delay time, were investigated for artificial pollutions. Increasing the laser energy intensity and selecting a suitable delay time could enhance the precision and relative spectral intensities of the elements. Additionally, reducing the particle size and increasing the density achieved the same results.

Keywords: laser-induced breakdown spectroscopy; surface pollution; high voltage insulators; quantitatively analysis

1. Introduction

The insulators were key equipment in transmission lines, in order to mechanically support conductor and give enough insulation space between conductor and tower. After being in operation for in a transmission line, an insulator (ceramic, glass or composite insulator) can accumulate a thick layer of pollutants on its surface due to different environmental factors. Under dry conditions, pollution was not harmful and had little effect on the safe service. However, soluble pollutants can be dissolved in water, forming a conductive water film on the surface of an insulator; this process results in the formation of conductive channels on the surface of the insulator, and in turn, reduces the pollution flashover voltage (PFV), thereby causing partial discharge, arc and even flash-over incidents [1]. Methods for detecting the pollution characteristics and pollution level of insulators have been studied for a long time. The Working Group 04 of Study Committee 33 (Over-voltage and Insulation Coordination) of the International Council on Large Electric Systems has recommended five methods for quantitatively characterizing pollution levels, including the equivalent salt deposit density (ESDD), surface conductivity, leakage current, PFV and pollution flashover gradient.

Pollution composition is complex and differs between environments. In nature, soluble pollutions are primarily conductive electrolytes, such as NaCl, KCl, $CaSO_4$, $CaCl_2$, Na_2SO_4, $NaNO_3$ and KNO_3; the main insoluble pollutions include SiO_2, C, Al_2O_3, $MgSO_4$, Fe_2O_3 and CaO [2,3]. Researchers have found that the pollution levels measured by ESDD differ from the actual values to a certain extent. As a result, the PFVs of artificial pollutions are lower than those of natural pollutions with the same ESDD. The PFV of the artificial pollution $CaSO_4$ is higher than that of the artificial pollution NaCl for the same ESDD. Additionally, for an artificial pollution mixture of NaCl and $CaSO_4$, the higher the $CaSO_4$ content, the higher the PFV is [4–7].

Currently, researchers also employ other indirect methods (e.g., light, sound and electricity) to determine the pollution levels. Hyperspectral imaging, microwave radiation theory, infrared and visible light information fusion, ultraviolet sensors, light detection sensors and acoustic emission technology have been employed to establish insulator pollution level prediction models [8–15]. With respect to the direct detection of pollution composition, aside from commonly used material composition analysis methods (e.g., ion emission spectroscopy techniques, including ion chromatography, X-ray diffraction and inductive coupling), very few researchers have examined online detection methods for insulator pollution composition. However, the compositional distribution of pollutions on surface of an insulator is often complex and heterogeneous. These factors present difficulties for evaluating pollution levels by indirect methods. Additionally, research results have demonstrated that pollution composition and material characteristics can affect the pollution flashover process, and may cause excessive or deficient insulation in insulation design [16–18]. The PFV is not only related to soluble salt composition, but also affected by insoluble substances in different mixtures [19].

To improve the accuracy and application of LIBS. The researchers studied various sample preparation techniques, such as dilution and using binding material, etc. By milling [20] and grinding, the particle size is reduced and the surface area is increased to make the sample more uniform. The smaller the particle size, the easier it is to evaporate and atomize in the plasma [21].

Laser-induced breakdown spectroscopy (LIBS) is a qualitative and quantitative analytical method based on pulse laser technology that examines the plasma atomic emission spectrum after exciting the sample [22], and it had higher sensitivity for light elements detection (H,Li,C,Si etc.), compared to EDS (or EDX) technique [23,24]. Currently, owing to the rapid development of this technique, the use of LIBS is widespread in the theoretical and experimental research of many fields, such as those of mineral products, archaeology, biomedicine and aerospace exploration [25–29]. In particular, LIBS is currently the only feasible technique in fields that require remote elemental analysis [26]. We have [30,31] evaluated the feasibility of using LIBS to achieve rapid, accurate, online monitoring of the ageing performance of silicone rubber and to determine the components (C, O, Fe and Si) that are closely related to the ageing state of silicone rubber. Combined with XPS technology, the linear calibration curves of these components were established. Based on the variation trend of element spectral intensity with depth, the depth of aging layer was obtained. However, compared with silicone rubber, pollution composition is more complex and varied. Therefore, when studying pollutions using LIBS, it is necessary to consider the effects of the properties of the pollutions and optimize the system parameters. In this study, the effects of various factors on the LIBS spectra of natural and artificial pollutions are examined and optimized system parameters are proposed.

2. Results and Discussion

2.1. Microanalysis and LIBS Testing of Pollutions Sampled

Figure 1 shows SEM images of the natural pollutions on the surface of the insulator at two randomly selected analytical points. As demonstrated in Figure 1, the pollutions at sampling point 1 were densely distributed and exhibited layer-by-layer stacking. In contrast, the pollutions at sampling point 2 were loosely arranged, and there were relatively large spaces between the pollutions. The process by which natural pollutions were adhered to the surface of an insulator is affected by the air flow in the environment. The uneven adhesive forces between particles and the surface of an insulator, due to the ageing of the insulator and the random interactions between particles, can result in uneven adherence of pollutions on the surfaces of adjacent insulators.

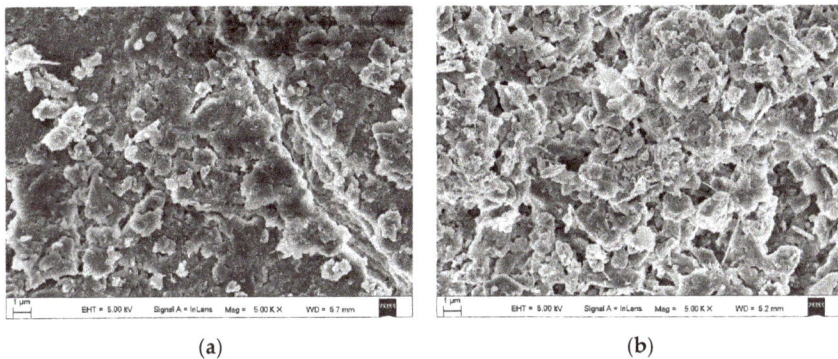

Figure 1. Scanning electron microscopy (SEM) images (5000×) of the natural pollutions on the surface of the insulator (**a**) Sampling point 1, (**b**) sampling point 2.

The EDS detector of the SEM was used to analyze the elements on the surface of the insulator. Table 1 summarizes the results. Very few elements were detected by EDS, and minimal Cl was detected. Natural pollutions often contain NaCl and KCl, which significantly affect pollution flashovers. The NaCl and KCl on the surface of the insulator may have been eliminated by dissolution and scouring as a result of dampening and rainfall. The EDS detector only analyzed the surface composition of the sample and consequently failed to detect the distributions of other common elements. Titanium ore is in the area of insulator operation, so there is high concentration of Ti in the pollution. Therefore, other methods were needed to determine the composition of pollutions on the surface of the insulator.

Table 1. Energy dispersive X-ray spectroscopy (EDS) analysis results for the composition of natural pollutions on the surface of the insulator.

Element	C	O	Na	Mg	Al	Si	Ti	Fe
wt/%	13.63	35.8	0.34	0.31	6.03	17.95	21.78	4.17

Figure 2 showed the LIBS spectrum of the natural pollutions. Table 2 showed the wavelength of typical spectral lines in Figure 2. In Figure 2, the wavelengths of the abscissa correspond to the emission intensities of various elements. Each element has multiple emission lines. In testing, a characteristic wavelength should be selected, and the type of element and relative spectral intensity, corresponding to the characteristic wavelength, should be determined [32]. Spectral intensities reflect the composition of the sample tested. As shown in Figure 2, Si, Ca, Al, C and Na had relatively high intensity, and this indicates that, agreeing well with the EDS area scan results, the natural pollutants had relatively high contents of these elements.

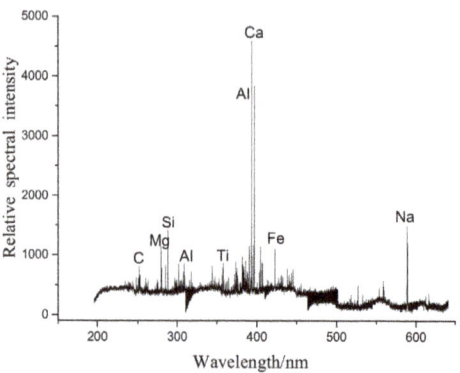

Figure 2. LIBS spectrum of the natural pollutions.

Table 2. Wavelength of typical spectral lines for pollutions element by LIBS.

Element	C I	Mg II	Si I	Al I	Ti I
Wavelength/nm	247.856	279.553	288.158	309.271	359.871
Element	Ca II	Al I	Fe I	Na I	Na I
Wavelength/nm	393.366	396.152	425.079	588.995	589.592

Trace amounts of Na and Mg were detected in the samples tested by LIBS, while the EDS area scans of the sample, Na and Mg were not detected in the ablation pits of the silicone rubber. Therefore, LIBS can not only achieve rapid, online detection of elements, but also help further reduce the detection limit of current composition testing and improve the accuracy of quantitative/qualitative compositional analysis.

2.2. Effects of Single-Pulse Laser Energy on the LIBS Signal

In LIBS, the depth of ablation craters depends on many factors, such as laser energy, ablation duration and material characteristics. The single-pulse laser energy has an impact on the ablation of pollutions on the surfaces of insulators. Ideally, a single-pulse laser beam only ablates the pollutions on the surface of an insulator but not the surface of the insulator itself. Figure 3 shows SEM images (200×) of the laser-ablated samples (labelled top and bottom). As demonstrated in Figure 3a,b, a focused laser beam produced an ellipsoidal ablation pit on the surface of each sample, which was related to the morphology of the focused laser beam. The sample bottom was taken from the surface of a silicone rubber insulator that had aged as a result of being in service for an extended period of time, and cracks differing in size were distributed on its surface. To further analyze the ablation effects of a single-pulse laser on the natural pollutions on the surface of the insulator, EDS area scans were performed on the natural pollutions and the ablation pits on the surfaces of the samples top and bottom to analyze the elemental compositions. The results showed that the typical characteristic elements (e.g., Na, Mg and Ti) were not detected in these samples by EDS. This observation suggests that the LIBS testing, with a single-pulse laser beam with an output energy of 110 mJ, was able to penetrate the relatively thin (micro-sized) pollution layers and ablated the pollutions at the point of action into laser plasma, thereby, exposing the substrate of the insulator.

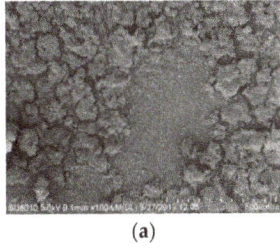 

(a) (b)

Figure 3. Distribution of the natural pollutions on the surface of the insulator (**a**) SEM image of the entire ablated sample top, (**b**) SEM image of the entire ablated sample bottom.

A laser energy increases within a certain range, the energy absorbed per unit target surface area increases, resulting in an increase in the spectral intensity of the sample. Once the increase in laser energy outside this range may result in self-absorption of or matrix effects on elements, which in turn, results in a decrease in intensity. In the experiment, artificial pollutions were prepared to determine the spectral intensities under various laser energies within a reasonable range. Figure 4. shows partial LIB spectra, obtained under various strengths of laser energies. As demonstrated in Figure 4, as the laser energy increased, the spectral intensities corresponding to different wavelengths increased by varying degrees. The spectral intensity of Al corresponding to a wavelength of 396.592 nm saturated prematurely. Therefore, while higher laser energy may improve the spectral intensity, extremely high laser energy outranging a certain range may interfere with the experiment.

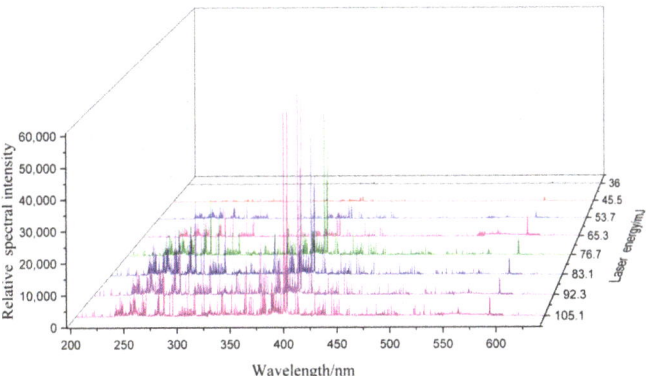

Figure 4. LIBS spectra within a certain band range under various laser energies.

This work done with the insulator that has been exposed to the elements. The LIBS method was described as follows: Five points on the surface of each sample were randomly selected. Each point was subjected to five continuous laser treatments. Figures 5 and 6 show the effects of laser energy density on the spectral intensity, and relative standard deviation (RSD) of various elements tested, respectively. The laser energy intensity was obtained by dividing the laser energy by the spot area. The diameter of laser focusing on sample surface was 0.8 mm. The spectral line intensity increased as the pulsed laser energy intensity increased. As demonstrated in Figure 6, as the spectral intensity increased, the RSDs of almost all the elements gradually decreased, suggesting that increasing laser energy intensity could effectively improve the repeatability of results. The RSD is related to the concentration and spectral line intensities of the sample and is affected by the spectral analysis conditions and instrument performance.

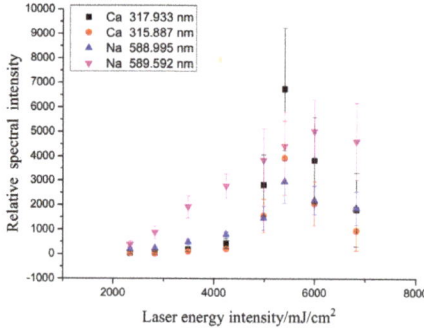

Figure 5. Effects of laser energy on the spectral intensities of the elements tested

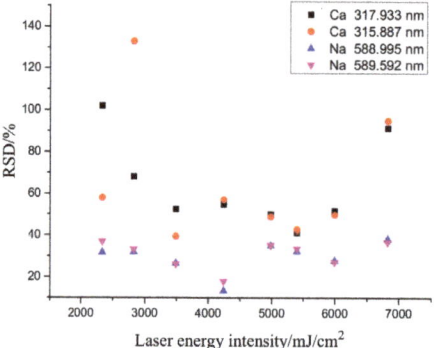

Figure 6. Relationship between the relative spectral intensity and measurement repeatability of the elements tested.

Additionally, an increase in laser energy intensity provided sufficient excitation energy for certain elements, causing intensity saturation or self-absorption effects and consequently decreasing the peak values. Meanwhile, owing to matrix effects, the increase in laser energy significantly interfered with the spectral information of other elements, leading to negative effects. Based on the SEM results for the pollutions subjected to LIBS testing, the laser energy was adjusted to approximately 80 mJ, corresponding to a laser ablation density of 3.814×10^{10} Watts/cm2. The ablation effects of the adjusted laser energy on the surface of the composite insulator were comparatively analyzed.

2.3. Selection of Delay Time

Figure 7 shows the trends of the spectral intensities within the same band range with the delay time. As demonstrated in Figure 7, as the delay time increased, the normalized spectral intensity corresponding to each wavelength significantly decreased. Using the average relative spectral intensity at a delay time of 0.5 µs as the baseline, the normalized relative spectral intensities were calculated by dividing the average relative spectral intensities at other delay times by the baseline. Figure 8 shows the results, the experimental results show that the trends of the spectral intensities of each element corresponding to various wavelengths were similar. Hence, only the spectral intensity of one element, corresponding to one wavelength, was selected for analysis.

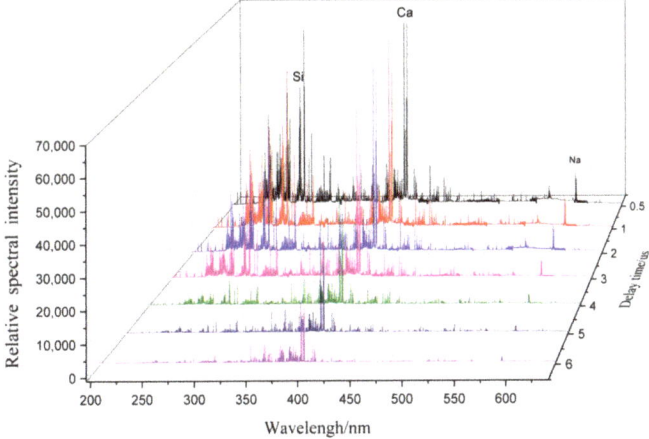

Figure 7. Changes in LIBS spectra within a certain band range with delay time.

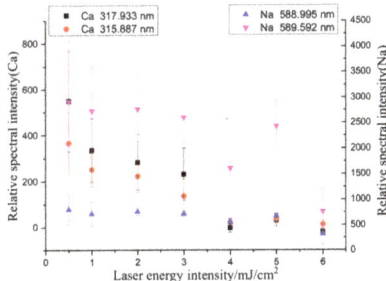

Figure 8. Effects of delay time on the relative spectral intensity of each element.

As demonstrated in Figure 8, the continuous background spectral process was not complete at a delay time of 0.5 μs. As the delay time increased, the spectral intensity of each element considerably decreased. Additionally, as the delay time increased, the RSD for Ca first slowly increased, and then gradually stabilized as shown in Figure 9. During the plasma cooling process, the collisions between ions and electrons continuously weakened, and consequently, the luminous intensities of energy released from the collisions and received by the spectrometer continuously decreased. In particular, as the delay time increased from 1 to 9 μs, the normalization ratio for Na fluctuated in the range of 0.3–0.45 because Na, being an alkali metal element prone to ionization, was completely ionized within 1 μs. As a result, as the measurement delay time increased, the number of Na ions received by the system decreased, resulting in a decrease in the measurement accuracy.

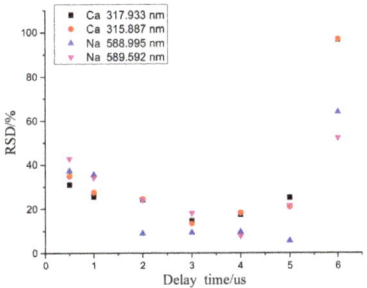

Figure 9. Effects of delay time on the repeatability of measurements on the elements tested.

Considering the relationships among the spectral intensity, RSD and delay time, a delay time ranging from 2 to 4 µs was selected as the optimum delay time range that led to a normalization ratio greater than 0.4 and an RSD less than 20%. A gate-width delay time of 3 µs was used in the subsequent experiment. When analyzing a particular element, a delay time range that leads to a normalization ratio greater than 0.5 and a minimum RSD should be selected.

2.4. Effects of Pollution Particle size and Density on LIBS Signal

Pollutions on the surface of an insulator in operation have complex and varied compositions (as shown in Figure 1). Pollution particles vary in size between different locations, and the gap density varies between pollutions. Inconsistent particle sizes and densities can both affect LIBS spectra.

First, the effects of pollution particle size on the LIBS spectral signal were studied. A Malvern Mastersizer 2000 laser particle-size analyzer was used to measure the particle size of the NaCl samples and kaolin clay [33]. A wet method was employed, and ethanol was used to dissolve the samples. Table 3 summarized the particle-size test results.

Table 3. Test results obtained using the laser particle-size analyser

Particle Size/µm	NaCl				Kaolin Clay
	<60 Mesh	60–100 Mesh	100–200 Mesh	200–300 Mesh	
Distribution/50%	342.3	240.763	140.694	60.914	5.346

The LIBS method was described as follows: Five points on the surface of each sample were randomly selected. Each point was subjected to five continuous laser treatments (frequency: 1 Hz). The relative spectral intensities of Na corresponding to wavelengths of 588.995 and 589.592 nm were extracted from the LIBS spectrum acquired for data analysis. The spectral intensities from 25 points on each sample were each divided by the background signal of the substrate, and the results were averaged (shown in Figure 10). As demonstrated in Figure 10, as the NaCl particle size decreased, the spectral intensities corresponding to the wavelengths of 588.995 and 589.592 nm gradually increased.

Based on the total area of XP-70, the mass of NaCl in each sample was determined. Each sample was compressed using a compression machine and subsequently subjected to LIBS testing. Figure 11 shows the changes in the relative spectral intensities of the Na in NaCl samples differing in particle size with NaCl concentration. When the NaCl particle size remained constant, the average relative spectral intensity of the two spectral lines of Na first increased and then decreased, as ESDD increased. For the NaCl samples with the same ESDD, the spectral intensity of Na was higher in the NaCl sample with a particle size of 60.914 µm than that in the NaCl sample with a particle size of 240.764 µm, exhibiting a trend similar to that in Figure 8.

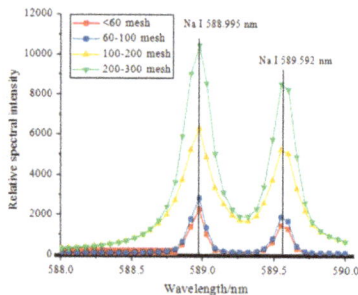

Figure 10. Spectral intensities of Na at 589 nm.

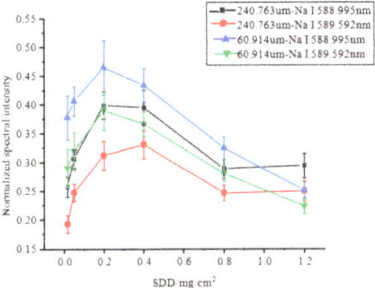

Figure 11. Changes in the spectral intensities of Na in NaCl samples differing in particle size with NaCl concentration.

Density is one of the variable properties of pollutions. Pollutions on the surfaces of insulators in different operating environments vary significantly in density. Thus, it is necessary to examine the effects of pollution density on LIBS spectral signals. Four identical artificial pollution samples, each consisting of kaolin clay (2 g) and NaCl (1%), were prepared. The four samples were compressed using a compression machine under compressive loads of 6, 9, 12 and 15 t. The compressed samples were subsequently subjected to LIBS testing to determine the relationship between compressive load and average relative spectral intensity (Figure 12). As demonstrated in Figure 12, as the density increased, the relative excited spectral intensities of the samples increased. This phenomenon can be explained by excited plasma plume dynamics. When the laser energy acts on the surface of a sample, the denser the surface of the sample is, the greater the impact of the laser pulse reverse shock wave is. Various types of particles jet from the target surface opposite the direction of the laser. The increase in the reverse jet velocity and intensity of various types of particles strengthens the collision ionization during the rapid expansion of the plasma, thereby, improving the atomic emission intensity.

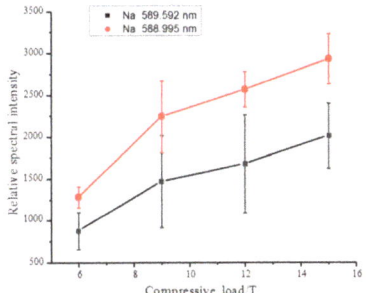

Figure 12. Effects of density on spectral intensity.

3. Experiments

A composite insulator chain (manufactured by Dongguan Gaoneng Industry Co., Ltd. in Dongguan, China) was collected from the N63 jumper of the 220-kV Dongguan–Kuihu line A. As shown in Figure 13, an insulator was cut from the centre of the insulator chain along the external surface of the core of the chain. A small piece (1 cm × 1 cm) was cut from a relatively dark-colored area of the insulator and subjected to scanning electron microscopy (SEM) and energy dispersive X-ray spectroscopy (EDS) analysis on a Zeiss Supra 55 SEM (manufactured by Carl Zeiss Co., Ltd. in Oberkochen, Germany) equipped with an Oxford X-Max 20 EDS detector (manufactured by Oxford Instruments Co., Ltd. in Oxford, Britain) to determine the content and distribution of the pollutions on the surface. A Leica EMACE 200 fully automatic low-vacuum coating system was used to coat the sample with Pt to improve its surface conductivity.

Figure 13. Schematic diagram of the insulators used in the experiment.

A LIBS system assembled by our research group was used in the experiment. This LIBS system consists of a Nimma-900 laser system (wavelength: 1,064 nm, pulse width: 10ns, output frequency: 1 Hz, and output energy: 110 mJ), focal spot diameter of approximately 80 μm, laser energy density of 2.1883×10^{11} Watts/cm^2, an Avantens spectrometer (available wavelength range: 200–650 nm) and a DG645 delay controller. The delay controller controls the interval between the output of the laser system and the acquisition of the spectrometer to effectively obtain an atomic emission spectrum evolved from a continuous background emission spectrum generated after plasma excitation under the action of the laser. The delay time was set to 3 μs in the experiment. The horizontal laser beam emitted by the laser system was reflected by a 45° mirror onto the vertical plane and focused by a convex lens onto the surface of the sample. The lens-to-sample distance was adjusted to position the sampling spot at the focal point of the convex lens. The spectral data acquired by the spectrometer were exported using the software Avasoft 8.8 (developed by Avantes Co., Ltd. in Apeldoorn, the Netherlands) and were subsequently processed.

In the experiment, kaolin clay was mixed with NaCl of different particle sizes at a 1:1 mass ratio. The shape of mixture is a circle with a diameter of 8 nm. Each mixture was compressed using a compression machine under compressive loads of 9t. and subsequently subjected to LIBS testing. Thus, the spectra of the Na in NaCl of different particle sizes were obtained. The NaCl particle size was determined using a laser particle size analyzer. In addition, the spectral intensity of NaCl obtained by LIBS was normalized to improve the analytical accuracy.

Spectrographic-grade NaCl samples (manufactured by Aladdin Industrial Co., Ltd. in Shanghai, China) were used in the experiment. Additionally, after sieving through <60, 60–100, 100–200 and 200–300 mesh stainless steel sieves, corresponding to particle sizes of >250, 150–250, 75–150 and 200–300 μm, respectively, NaCl samples of four different particle sizes were obtained (50 g of each type).

4. Conclusions

In this study, the microregional characteristics and element distributions of natural pollutions were analyzed with LIBS. The conclusions derived from this study are summarized as follows:

(1) Natural pollutions were obtained from the surface of an composite insulator from a 220kV transmission line. Through EDS, the main elements (Na, Mg, Si, Fe, O and C) composing the pollution sample were detected, which are common elements in natural pollutions. LIBS detected compositional elements of the pollution sample, meanwhile EDS failed to detect, thus, effectively reducing the element detection limit.

(2) A 110-mJ laser pulse was sufficient to penetrate the artificial pollutions on the surface of the insulator. With the accumulation of pulses, the relative spectral intensities of the common pollution elements on the LIBS spectrum gradually decreased.

(3) Artificial pollutions were prepared. The effects of the LIBS delay time and laser energy on the spectral signals were examined. The results showed that selecting a suitable delay time could improve the repeatability of data detection. In this study, the delay time was set to 3 μs. An increase in the laser energy increased the relative spectral intensity and RSD of each element. A suitable laser energy must be selected so that the laser does not harm the insulator substrate.

In this study, the laser energy was set to 80 mJ, corresponding to a laser energy ablation density of 3.814×10^{10} Watts/cm^2.

(4) The effects of the pollution properties (particle size and density) on the spectral signals were analyzed. A decrease in the particle size and an increase in the density of the sample both, improved the relative spectral intensities of the elements tested.

Author Contributions: Conceptualization and formal analysis, X.W. (Xinwei Wang) and S.L.; investigation, X.Q.; resources, X.W. (Xinwei Wang) and T.W.; data curation, S.L. and T.W.; writing—original draft preparation, X.Q.; writing—review and editing, X.W. (Xilin Wang) and Z.J.; supervision, X.W. (Xilin Wang) and Z.J. All authors have read and agreed to the published version of the manuscript.

Funding: This research was funded by National Natural Science Foundation of China (51607101), Science and technology projects of Shanxi Electric Power Research Institute (SGSXDK00SPJS1900162), and the Guangzhou Science and Technology Plan (201707020044).

Conflicts of Interest: The authors declare no conflict of interest.

References

1. Guan, Z.C. *External Insulation of Insulator and Power Transmission Equipment*; Tsinghua University Press: Beijing, China, 2006.
2. Ramos, G.N.; Campillo, M.T.R.; Naito, K. A study on the characteristics of various conductive contaminants accumulated on high voltage insulators. *IEEE Trans. Power Deliv.* **1993**, *8*, 1842–1850. [CrossRef]
3. Takasu, K.; Shindo, T.; Arai, N. Natural contamination test of insulators with DC voltage energization at inland areas. *IEEE Trans. Power Deliv.* **1988**, *3*, 1847–1853. [CrossRef]
4. Moula, B.; Mekhaldi, A.; Teguar, M.; Haddad, A. Characterization of discharges on non-uniformly polluted glass surfaces using a wavelet transform approach. *IEEE Trans. Dielectr. Electr. Insul.* **2013**, *20*, 1457–1466. [CrossRef]
5. Seta, T.; Nagai, K.; Naito, K.; Hasegawa, Y. Studies on performance of contaminated insulators energized with dc voltage. *IEEE Trans. Power Appar. Syst.* **1981**, *100*, 518–527. [CrossRef]
6. Jolly, D.C.; Poole, C.D. Flashover of contaminated insulators with cylindrical symmetry under DC conditions. *IEEE Trans. Dielectr. Electr. Insul.* **1979**, *14*, 77–84. [CrossRef]
7. Dey, D.; Chakravorti, S.; Chatterjee, B.; Chakravorti, S. Low- complexity leakage current acquisition system for transmission line insulators employing GSM voice channel. *Electr. Lett.* **2015**, *51*, 1538–1540.
8. Ahmadi, J.I.; Shayegani, A.A.A.; Mohseni, H. Leakage current analysis of polymeric insulators under uniform and non-uniform pollution conditions. *IET Gener. Transm. Distrib.* **2017**, *11*, 2947–2957. [CrossRef]
9. Jia, Z.D.; Chen, C.; Wang, X.L.; Lu, H.; Yang, C.; Li, T. Leakage current analysis on RTV coated porcelain insulators during long term fog experiments. *IEEE Trans. Dielectr. Electr. Insul.* **2014**, *21*, 1547–1553.
10. Jin, L.J.; Ai, J.Y.; Tian, Z.R.; Zhang, Y. Detection of polluted insulators using the information fusion of multispectral images. *IEEE Trans. Dielectr. Electr. Insul.* **2017**, *24*, 3530–3538. [CrossRef]
11. Yang, H.; Zhang, Q.G.; Pang, L.; Gou, X.; Yang, X.; Zhao, J.; Zhou, J. Study of the AC arc discharge characteristics over polluted insulation surface using optical emission spectroscopy. *IEEE Trans. Dielectr. Electr. Insul.* **2015**, *22*, 3226–3233. [CrossRef]
12. Ferreira, T.V.; André, D.G.; Costa, E.G.D. Ultrasound and Artificial Intelligence Applied to the Pollution Estimation in Insulations. *IEEE Trans. Power Deliv.* **2012**, *27*, 583–589. [CrossRef]
13. Li, H.L.; Wen, X.S.; Shu, N.Q.; Pei, C. Application of Acoustic Emission Technology on Monitoring of Polluted Insulator Discharge'. In Proceedings of the 2009 Asia-Pacific Power and Energy Engineering Conference, Wuhan, China, 27–31 March 2009; pp. 1–4.
14. Yin, J.; Lu, Y.; Gong, Z.; Jiang, Y.; Yao, J. Edge detection of high-voltage porcelain insulators in infrared image using dual parity morphological gradients. *IEEE Access* **2019**, *7*, 32728–32734. [CrossRef]
15. Li, L.; Li, Y.Q.; Lu, M.; Liu, Z.; Wang, C.; Lv, Z. Quantification and comparison of insulator pollution characteristics based on normality of relative contamination values. *IEEE Trans. Dielectr. Electr. Insul.* **2016**, *23*, 965–973. [CrossRef]

16. Zhang, Z.J.; Zhang, D.D.; You, J.W.; Zhao, J.; Jiang, X.; Hu, J. Study on the DC flashover performance of various types of insulators with fan-shaped non-uniform pollution. *IEEE Trans. Power Deliv.* **2015**, *30*, 1871–1879. [CrossRef]
17. Williams, L.J.; Kim, J.H.; Kim, Y.B.; Arai, N.; Shimoda, O.; Holte, K. Contaminated Insulators-Chemical Dependence of Flashover Voltages and Salt Migration. *IEEE Trans. Power Appar. Syst.* **1974**, *93*, 1572–1580. [CrossRef]
18. Jiang, X.L.; Wang, S.H.; Zhang, Z.J.; Hu, J.; Hu, Q. Investigation of flashover voltage and non-uniform pollution correction coefficient of short samples of composite insulator intended for ±800kV UHVDC. *IEEE Trans. Dielectr. Electr. Insul.* **2010**, *17*, 71–80. [CrossRef]
19. Zhang, Z.J.; Zhang, W.; You, J.W.; Jiang, X.; Zhang, D.; Bi, M.; Wu, B.; Wu, J. Influence factors in contamination process of XP-160 insulators based on computational fluid mechanics'. *IET Gener. Transm. Distrib.* **2016**, *10*, 4140–4148. [CrossRef]
20. Gomes, M.D.S.; Santos, D.; Nunes, L.C.; De Carvalho, G.G.A.; Leme, F.D.O.; Krug, F.J. Evaluation of grinding methods for pellets preparation aiming at the analysis of plant materials by laser induced breakdown spectrometry. *Talanta* **2011**, *85*, 1744–1750. [CrossRef]
21. Jantzi, S.C.; Motto-Ros, V.; Trichard, F.; Markushin, Y.; Melikechi, N.; De Giacomo, A. Sample treatment and preparation for laser-induced breakdown spectroscopy. *Spectrochim. Acta Part B At. Spectrosc.* **2016**, *115*, 52–63. [CrossRef]
22. Cremers, D.A.; Radziemski, L.J. *Handbook of Laser-Induced Breakdown Spectroscopy*; John Wiley & Sons, Ltd.: Hoboken, NJ, USA, 2006.
23. Gornushkin, I.B.; Smith, B.W.; Nasajpour, H.; Winefordner, J.D. Identification of solid materials by correlation analysis using a microscopic laser-induced plasma spectrometer. *Anal. Chem.* **1999**, *71*, 5157–5164. [CrossRef]
24. Pershin, S.M. Laser-induced breakdown spectroscopy for three-dimensional elemental mapping of composite materials synthesized by additive technologies. *Appl. Opt.* **2017**, *56*, 9698–9705.
25. Lanza, N.L.; Clegg, S.M.; Wiens, R.C.; McInroy, R.E.; Newsom, H.E.; Deans, M.D. Examining natural rock varnish and weathering rinds with laser-induced breakdown spectroscopy for application to ChemCam on Mars. *Appl. Opt.* **2012**, *51*, B74–B82. [CrossRef] [PubMed]
26. Ji, G.L.; Ye, P.C.; Shi, Y.J.; Yuan, L.; Chen, X.; Yuan, M.; Zhu, D.; Chen, X.; Hu, X.; Jiang, J. Laser-induced breakdown spectroscopy for rapid discrimination of heavy metal contaminated seafood tegillarca granosa'. *Sensors* **2017**, *17*, 2655. [CrossRef] [PubMed]
27. Stratis, D.N.; Eland, K.L.; Angel, S.M. Enhancement of aluminum, titanium, and iron in glass using pre-ablation spark dual-pulse LIBS. *Appl. Spectrosc.* **2000**, *54*, 1719–1726. [CrossRef]
28. Sathiesh, K.V.; Vasa, N.J.; Sarathi, R. Remote surface pollution measurement by adopting a variable stand-off distance based laser induced spectroscopy technique. *J. Phys. D Appl. Phys.* **2015**, *48*, 435504. [CrossRef]
29. Praher, B.; Palleschi, V.; Viskup, R.; Heitz, J.; Pedarnig, J. Calibration free laser-induced breakdown spectroscopy of oxide materials'. *Spectrochim. Acta B* **2010**, *65*, 671–679. [CrossRef]
30. Wang, X.L.; Hong, X.; Wang, H.; Chen, C.; Zhao, C.; Jia, Z.; Wang, L.; Zou, L. Analysis of the silicone polymer surface aging profile with laser-induced breakdown spectroscopy. *J. Phys. D Appl. Phys.* **2017**, *50*, 415601. [CrossRef]
31. Wang, X.L.; Hong, X.; Chen, P.; Zhao, C.; Jia, Z.; Wang, L.; Lv, Q.; Huang, R.; Liu, S. In-situ and quantitative analysis of aged silicone rubber materials with laser-induced breakdown spectroscopy. *High Volt.* **2018**, *3*, 140–146.
32. National Institute of Standards and Technology USA. Department of Technology Database. Available online: http://www.physics.nist.gov (accessed on 27 June 2019).
33. International Electrotechnical Commission. *EC/TR2 61245:1993: Artificial Pollution Tests on High-Voltage Insulators to be used on d.c. Systems*; International Electrotechnical Commission: Geneva, Switzerland, 1993.

Sample Availability: The composite insulator chain and samples of the NaCl in different particle size are available from the authors.

© 2020 by the authors. Licensee MDPI, Basel, Switzerland. This article is an open access article distributed under the terms and conditions of the Creative Commons Attribution (CC BY) license (http://creativecommons.org/licenses/by/4.0/).

Article

Development of a Direct Competitive ELISA Kit for Detecting Deoxynivalenol Contamination in Wheat

Li Han [1,2,†], Yue-Tao Li [2,†], Jin-Qing Jiang [2], Ren-Feng Li [2], Guo-Ying Fan [2], Jun-Mei Lv [2], Ye Zhou [2], Wen-Ju Zhang [1,*] and Zi-Liang Wang [2,*]

1. College of Animal Science and Technology, Shihezi University, Shihezi 832000, China; HanLi3909@163.com
2. College of Animal Science and Veterinary Medicine, Henan Institute of Science and Technology, Xinxiang 453003, China; liyuetao2019@163.com (Y.-T.L.); jjq5678@126.com (J.-Q.J.); lirenfeng2019@163.com (R.-F.L.); fanguoy@163.com (G.-Y.F.); junmeirosa@163.com (J.-M.L.); YeZhou9502@163.com (Y.Z.)
* Correspondence: zwj@shzu.edu.cn (W.-J.Z.); wangziliang1966@163.com (Z.-L.W.)
† These authors contributed equally to this article.

Academic Editor: Clinio Locatelli
Received: 4 November 2019; Accepted: 18 December 2019; Published: 22 December 2019

Abstract: This study was conducted to develop a self-assembled direct competitive enzyme-linked immunosorbent assay (dcELISA) kit for the detection of deoxynivalenol (DON) in food and feed grains. Based on the preparation of anti-DON monoclonal antibodies, we established a standard curve with dcELISA and optimized the detection conditions. The performance of the kit was evaluated by comparison with high-performance liquid chromatography (HPLC). The minimum detection limit of DON with the kit was 0.62 ng/mL, the linear range was from 1.0 to 113.24 ng/mL and the half-maximal inhibition concentration (IC_{50}) was 6.61 ng/mL in the working buffer; there was a limit of detection (LOD) of 62 ng/g, and the detection range was from 100 to 11324 ng/g in authentic agricultural samples. We examined four samples of wheat bran, wheat flour, corn flour and corn for DON recovery. The average recovery was in the range of 77.1% to 107.0%, and the relative standard deviation (RSD) ranged from 4.2% to 11.9%. In addition, the kit has the advantages of high specificity, good stability, a long effective life and negligible sample matrix interference. Finally, wheat samples from farms in the six provinces of Henan, Anhui, Hebei, Shandong, Jiangsu and Gansu in China were analyzed by the kit. A total of 30 samples were randomly checked (five samples in each province), and the results were in good agreement with the standardized HPLC method. These tests showed that the dcELISA kit had good performance and met relevant technical requirements, and it had the characteristics of accuracy, reliability, convenience and high-throughput screening for DON detection. Therefore, the developed kit is suitable for rapid screening of DON in marketed products.

Keywords: deoxynivalenol; dcELISA kit; performance measurement; development

1. Introduction

Deoxynivalenol (DON), also known as vomitoxin, is a highly toxic secondary metabolite produced by *Fusarium graminearum* and *Fusarium culmorum*; DON belongs to the B-group of trichothecenes and widely exists in various agricultural products, food, and animal feed, especially in wheat, maize, and other cereal crops [1–4]. DON readily acts as an animal antifeedant and shows immunotoxicity, organ toxicity, inhibition of protein synthesis, and teratogenicity. These symptoms are closely related to immune suppression, Keshan disease, oesophageal cancer and other diseases [5–7]. Moreover, DON is heat-stable, and general cooking and processing cannot destroy its toxicity. Young et al. [8] found that grain processed into pet food still contained DON. Therefore, DON pollution poses a great threat to human and livestock health and has attracted the attention of countries around the world [9]. At present, at least 100 countries have mandatory limits on DON levels in food and feed. In view of its serious toxic effects, in the preliminary draft of the DON maximum levels (MLs), the Codex Alimentarius Food (FAO) Committee recommended the following limits: 2 mg/kg in unprocessed cereals, 1 mg/kg in semi-processed products using wheat, corn and barley as raw materials, and 0.5 mg/kg in cereals for infants and young children [10,11]. In China, the ML of DON in maize, wheat, and their products is regulated at 1 mg/kg [12,13]. The molecular formula of DON is $C_{15}H_{20}O_6$, and its molecular weight is 296.32 [14]. Its structure is shown in Figure 1:

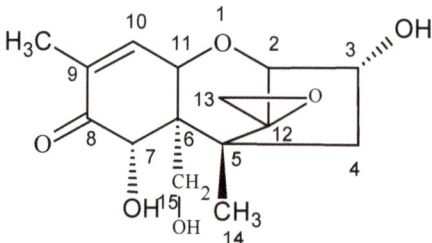

Figure 1. Molecular structure of deoxynivalenol (DON).

At present, the main physical and chemical methods for detecting DON contamination in food and feed are thin-layer chromatography (TLC), high-performance liquid chromatography (HPLC), gas chromatography (GC), mass spectrometry (MS), gas chromatography-mass spectrometry (GC-MS), high-performance liquid chromatography–mass spectrometry (HPLC–MS), high-performance liquid chromatography–tandem mass spectrometry (HPLC–MS/MS), and others [15–17]. These methods have high precision and sensitivity, but the sample pretreatment is rigorous, the instruments are expensive, the detection range is small, and, as the analysts often need special training, the cost is high. These methods are only suitable for large enterprises, scientific research institutes, or testing institutions that require high detection sensitivity, and they are not suitable for the demand of DON pollution detection in the feed industry. Therefore, increasing attention has been paid to the simple, rapid, sensitive, low-cost enzyme-linked immunosorbent assay (ELISA), which is suitable for large-scale sample screening. For example, the traditional immunoassay (ELISA) [18], chemiluminescence enzyme immunoassay (CLEIA) [19], fluorescence polarization immunoassay (FPIA) [20], time-resolved fluorescence immunoassay (TRFIA) [21], colloidal gold immunochromatography (GICA) [22,23], surface plasmon resonance (SPR) immunoassay [24], silver-stained GICA [25], nanobody-based ELISA [26], and immunosensor, among others, can be used to detect DON. Therefore, due to the prevalence of DON contamination and the large number of samples that need to be analyzed, ELISA kits have been considered a suitable detection tool, and their development and application has grown rapidly in recent years because they do not need special instruments and equipment, are suitable for the field and are suitable for high-throughput screening.

The purpose of this experiment is to assemble and optimize a new DON dcELISA kit. The performance of the ELISA kit was tested, and its accuracy was verified by HPLC, which laid a foundation for the development of ELISA kits with high sensitivity, specificity, and good quantification suitable for screening a large number of DON-contaminated samples.

2. Materials and Methods

2.1. Reagents and Materials

The standards of DON, 3-Ac-DON, 15-Ac-DON, Nivalenol (NIV), Fusarenon-X, T-2 toxin, Zearalenone (ZEN), and Aflatoxin B1 (AFB1) were purchased from Sigma-Aldrich Co., Ltd. (Augsburg, Germany). Bovine serum albumin (BSA), chicken ovalbumin (OVA), *N,N′*-carbonyldiimidazole (CDI), anhydrous tetrahydrofuran (THF), *N,N*-dimethylformamide (DMF), horseradish peroxidase (HRP),

1-ethyl-3-(3-dimethylamino)propyl) carbodiimide hydrochloride (EDC), Freund's complete adjuvant (FCA), and Freund's incomplete adjuvant (FIA) were provided by Pierce. PEG-1500 (polyethylene glycol) was purchased from Roche. GaMIgG was purchased from Huamei Biotechnology Company (Shanghai, China). In addition, 96-well microtiter plates as well as 24-well and 96-well cell culture plates were purchased from Iwaki Co., Ltd. (Dalian, China); 3,3,5,5-Tetramethylbenzidine (TMB), phenacetin and urea peroxide were purchased from Sigma. Foetal bovine serum (FBS) was purchased from Gibco. Female Balb/c mice (6 to 8 weeks old) were provided by Beijing SPF Biotech Co., Ltd. (Beijing, China) and were raised under strict control in our laboratory animal house.

Phosphate-buffered saline (PBS), carbonate-buffered saline (CBS), washing buffer (PBST, PBS containing 0.05% Tween-20), blocking buffer (SPBST, PBST containing 5% goat serum), color substrate solution (TMB), stopping solution (2 M H_2SO_4), Glucose sodium chloride potassium chloride solution (GNK), complete medium, Hypoxantin Aminopterin and Thymidin (HAT) medium, Hypoxantin and Thymidin (HT) medium, were all made in-house in our laboratory.

A Galaxy S-type CO_2 cell incubator was purchased from Biotech. A Multiskan MK3 microplate reader was purchased from Thermo (Waltham, Ma, USA) and used for 450 nm absorbance measurements. An inverted MIC 00949 microscope was purchased from Nikon Corporation. A DK-8D water bath was provided by Yiheng Instrument Co., Ltd. (Shanghai, China). A BS124S electronic balance was purchased from the German Sartorius Group. Purified water was prepared using a Milli-Q purification system (Millipore Corporation, Bedford, MA, USA). An A11 basic analytical mill was provided by IKA (Staufen, Germany). A Legend Micro 17 centrifuge was provided by Thermo (Waltham, MA, USA). Glass microfiber filter paper was purchased from Whatman (Maidstone, UK). The reliability of the ELISA kit was confirmed using an Agilent 1260 HPLC equipped with a diode array detector (DAD) (Agilent Technologies, Wilmington, DC, USA).

2.2. Preparation of the Antigen and Anti-DON Monoclonal Antibody (mAb)

According to the molecular structure of DON, the artificial antigen DON-BSA was synthesized by the carbonyl diimidazole (CDI) procedure outlined in a previously published method by Maragos et al. [14], with slight modifications. The synthesis of coated DON-OVA was improved by referring to the method of Li et al. [27]. DON was derivatized by maleic anhydride, and then the hapten was coupled with OVA to coat the original DON-OVA by implementation of the carbodiimide (EDC) procedure. Preparation of anti-DON mAb was achieved with classical hybridoma technology [28]. After obtaining DON mAb hybridoma cell lines, this experiment adopted an in vivo induced ascites method [29] to mass produce DON mAb, which was then purified from ascites by an octanoic acid/ammonium sulfate precipitation method [30]. The DON mAb was then stored at −20 °C until the dcELISA kit was assembled.

2.3. Development of the DON dcELISA Kit

We prepared the enzyme-labeled hapten (horseradish peroxidise-DON) and determined its working concentration as follows.

The enzyme-labeled hapten (HRP-DON) was prepared by a carbonyl diimidazole (CDI) method. DON standard (5 mg) was dissolved in 1 mL THF, 60 mg CDI was added, and the reaction proceeded for 4 h in a dry environment at 70 °C. The solvent of the reaction products was evaporated, and 500 µL DMF was added to the remaining products and completely dissolved. Then, 2 mg of HRP was added dropwise (the HRP was dissolved in 2 mL 0.01 mol/L pH 7.4 PBS solution) and stirred for 24 h at 4 °C in the dark. The reaction products were dialyzed in PBS for 72 h, the fluids were replaced 9 times during dialysis, and the dialysate, which was the enzyme-labeled hapten HRP-DON, was collected.

It is well known that the working concentration of the coated antigen and antibody is the key to determine the sensitivity of an ELISA kit. To determine the optimum dilution of RaMIgG, anti-DON mAb, and HRP-DON, chessboard titration tests were carried out. HRP-DON was added to 50% glycerol and stored at −20 °C.

2.4. Components of the ELISA Kit

The optimum conditions of the kit were very important for improving detection technology. The components and parameters of the kit are shown in Table 1 [31]:

Table 1. Components and parameters of the direct competitive enzyme-linked immunosorbent assay (dcELISA) kit.

Number	Composition	Quantity	Unit	Parameters
1	ELISA microplates	1	board	coated 96-well transparent microplates
2	Mab solution	1	tube	6 mL (concentration of 1:6400 diluted in PBS)
3	DON standard	1	tube	1 mg/mL
4	HRP-DON	1	tube	3 mL (concentration of 1:800 diluted in PBS)
5	1 × Working buffer	1	bottle	50 mL (5% methanol, 0.5 mol/L Na$^+$, pH 7.4 in PBS)
6	10 × Washing buffer	1	bottle	50 mL (10 × PBST, pH 7.4)
7	Color substrate buffer	1	bottle	15 mL (0.4 mmol/L TMB and 3 mmol/L H$_2$O$_2$ diluted in citrate buffer, pH 5.0)
8	Stop buffer	1	bottle	10 mL (2 mol/L H$_2$SO$_4$ diluted in H$_2$O)

2.5. Establishment of the Kit Standard Curve

The standard curve was established by dcELISA. The inhibition rate B/B_0 of a series of concentrations of DON standards against DON mAb was taken as the ordinate, and the logarithmic value of a series of concentrations of DON standards was taken as the abscissa. The standard curve was analyzed and fitted using Origin Program 7.0 software (OriginLab Co., Northampton, MA, USA), and the linear regression was established. The theoretical detection limit and linear detection range of the kit were calculated by the regression equation.

2.6. Pretreatment of Samples

The wheat samples came from farms in six Chinese provinces: Henan, Anhui, Hebei, Shandong, Jiangsu and Gansu. A total of 30 samples were randomly checked (5 samples from each province). After the samples were ground, 5 g of each sample was accurately weighed (accurate to 0.01 g) and placed in a bottle. Distilled water (25 mL) was added, and extracted by sonication for 10 min. The mixture was evenly mixed for a few minutes. The supernatant was centrifuged at 8000 rpm/min for 5 min. Finally, 500 µL of the supernatant was added to 500 µL of the sample diluent, which is the extract solution of the sample to be tested. In addition, the pH of the sample extract was adjusted from 6 to 8. If needed, samples were diluted with the working buffer before being analyzed with the kit.

2.7. Operating Procedure of the Kit

(1) Addition of anti-DON mAb: anti-DON mAb at a working concentration was added (50 µL/well), set as the negative and blank control, incubated for 15 min at 37 °C, and then washed.
(2) Addition of HRP-DON and the sample to be tested: HRP-DON at a working concentration was added (50 µL/well), the sample to be tested was added at the same volume, the plate was incubated for 25 min at 37 °C, and then washed.
(3) Coloration: the TMB-containing color substrate solution (50 µL/well) was added, and the plate was placed in the dark for 5 min at room temperature (RT).
(4) Termination: a 2 mol/L H$_2$SO$_4$ termination solution was added (50 µL/well).
(5) Measurement with the microplate reader: the absorbance was measured at 450 nm, and the inhibition rate was calculated.

2.8. Characteristics of the DON dcELISA Kit

2.8.1. Sensitivity Determination

According to the method of Hayashi et al. [32], the sensitivity of competitive ELISA is $B/B_0\% = 83.3\%$; the sensitivity of the kit was calculated according to the standard curve regression equation, and the detection limit was also determined.

2.8.2. Accuracy and Precision Determination

In this study, the accuracy and precision of the kit were determined with recovery experiments and expressed as recovery (%) and relative standard deviation (RSD%), respectively. The wheat bran, wheat flour, maize flour, and maize were first treated with 1% Na_2CO_3 for detoxification [33]. Then, 5 g of each sample was spiked with DON at 200, 500 and 1000 ng/g and stirred for 2 h at room temperature (RT). Next, the spiked samples were added to 10 mL of working buffer containing 20% methanol, and extracted by sonication for 10 min. The supernatant was centrifuged at 8,000 rpm/min for 5 min. Finally, 500 μL of the supernatant was added to 500 μL of the working buffer, which is the extract solution of the sample to be tested. Then, each sample was tested three times, and the recovery (%) and RSD% were calculated:

$$\text{Recovery (\%)} = \text{the measured value/the actual added value} \times 100\% \qquad (1)$$

$$\text{RSD (\%)} = \text{SD (standard deviation)} / \overline{X} \text{ (mean value)} \times 100\% \qquad (2)$$

2.8.3. Specificity Determination

The specificity of the cross-reactions between the kit and other mycotoxins was evaluated, and the formula of cross-reaction rate (CR%) is [34]:

$$CR\ (\%) = [IC_{50}\ (DON)/IC_{50}\ (\text{Structural Analogue})] \times 100\% \qquad (3)$$

2.8.4. Stability Determination

The stability of the kit was evaluated by the changes in B_0 (the value of absorbance without the DON standard) and B/B_0 (%) (the ratio value of absorbance with 5 ng/mL DON and without the DON standard) during storage (2 to 8 °C).

2.8.5. Matrix Effect Determination

To analyze the effect of the sample matrix on the sensitivity of the kit, the DON standard solution was dissolved in four samples of wheat bran, wheat flour, corn flour, and corn. These samples were then diluted with sample diluent, and the curve was generated according to the operation of the kit.

2.9. Confirmation of the DON dcELISA Kit with HPLC

The wheat samples from farms in the six provinces of Henan, Anhui, Hebei, Shandong, Jiangsu and Gansu in China were tested using the assembled DON dcELISA kit and HPLC. A total of 30 samples were randomly checked (5 samples from each province), and the correlation between the kit and HPLC was evaluated by comparing the results of detection [35]. Sample extraction and HPLC analysis were performed according to the method of the national standard of China GB5009.111-2016 [36], with slight modifications. After the samples were ground, 5 g of each sample was accurately weighed (accurate to 0.01 g) and placed in a clean and capped wide-mouth bottle. Twenty-five milliliters of acetonitrile-H_2O (20:80, v/v) and 2 g polyethylene glycol were added. The bottle was capped and extracted by sonication for 30 min. The mixture was evenly mixed for a few minutes. The samples were centrifuged at 6000 rpm/min for 10 min. Finally, the supernatants were filtered through glass microfiber filters to clarify the extract solution of the sample to be tested. Then, the supernatants were purified through DON immunoaffinity columns. The extracted phases were collected and analyzed by HPLC. The HPLC analysis was performed using an Agilent 1260 HPLC equipped with a diode array detector (DAD). Separation was performed on a C18 liquid chromatographic column (150 mm × 4.6 mm × 5 μm) or equivalent, the mobile phase was methanol:water (20:80, v/v), the flow rate was 0.8 mL/min, the column temperature was 35 °C, the injection volume was 50 μL, and the detection wavelength was 218 nm.

3. Results

3.1. Development of the DON dcELISA Kit

For the determination of the working concentrations of RaMIgG, anti-DON mAb and HRP-DON using the chessboard titration tests, ELISA microplates were coated with 10 ng/mL RaMIgG. The working concentrations of the anti-DON mAb and HRP-DON were determined as 1:6400 (1.56 ng/mL) and 1:800 (28.5 ng/mL), respectively, when the value of B_0 reached 1.0.

The key parameters were studied to guarantee the ideal sensitivity and performance of the kit for detecting DON. Under the criteria of a higher value of B_0/half-maximal inhibition concentration (IC_{50}) and lower value of IC_{50}, the working buffer, which could greatly affect the sensitivity of the kit, was adjusted. Finally, 5% methanol, 0.5 mol/L Na^+, and pH 7.4 in the working buffer were selected as the optimal working buffer for the kit (Table 2).

Table 2. Key parameters for the proposed kit.

Factor	Parameter	Factor	Parameter
RaMIgG	10 ng/mL	Methanol (v/v, %)	5
anti-DON mAb	1:6400 (1.56 ng/mL)	Na^+ (mol/L)	0.5
HRP-DON	1:800 (28.5 ng/mL)	pH	7.4

3.2. Generating and Fitting the Standard Curve of the Kit

The standard curve of the kit is shown in Figure 2. By analyzing the curve, the regression equation $y = -32.433x + 76.608$, correlation coefficient $R^2 = 0.972$, and $IC_{50} = 6.61$ ng/mL was obtained, and the detection range (IC_{10} to IC_{80}) was 1.0 to 113.24 ng/g.

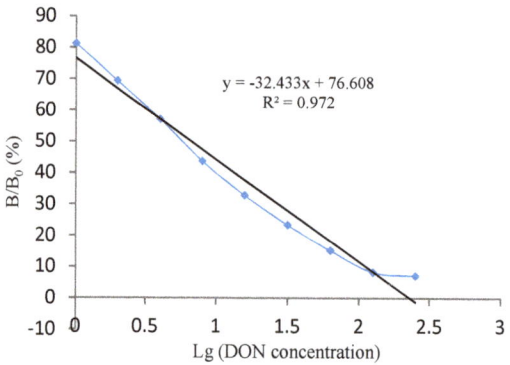

Figure 2. Calibration curve of the dcELISA kit.

3.3. Performance Measurements of the Kit

3.3.1. Sensitivity Determination

When $B/B_0 = 83.3\%$, the corresponding DON concentration was 0.62 ng/g, indicating a sensitivity of 0.62 ng/g, which was obtained by substituting the B/B_0 value into the standard curve regression equation. However, considering the need for positive detection and the error of user operation, the detection limit of the competitive ELISA kit was determined to be 1.0 ng/g.

3.3.2. Accuracy and Precision Measurement

Table 3 shows the four feed samples of wheat bran, wheat flour, corn flour, and corn with the recoveries. The average recovery was in the range of 77.1% to 107.0%, and the RSD ranged from 4.2% to 11.9%. The dcELISA kit meets the requirements of national accuracy and precision, indicating that the kit can be used for the detection of actual samples.

Table 3. Recoveries of DON in different samples by the dcELISA kit ($n = 3$).

Samples	Spiked (ng/g)	Mean Recovery ± SD (%)	RSD (%)
Wheat bran	200	89.2 ± 6.2	7
	500	88.1 ± 5.7	6.5
	1000	79.4 ± 7.5	9.4
Wheat flour	200	77.1 ± 9.2	11.9
	500	81.7 ± 5.6	6.8
	1000	96.5 ± 4.1	4.2
Corn meal	200	104.4 ± 5.8	5.5
	500	96.4 ± 6.3	6.5
	1000	107.0 ± 7.6	7.1
Corn	200	103.7 ± 4.6	4.4
	500	95.0 ± 5.3	5.6
	1000	98.4 ± 7.3	7.4

3.3.3. Specificity Determination

Table 4 shows that the cross-reactions between the kit and other mycotoxins were negligible. The cross-reaction rate with 3-Ac-DON was 4.7% and that with other mycotoxins was less than 0.2%, indicating that the kit has high specificity.

Table 4. Cross-reactivity of the DON dcELISA kit with other related mycotoxins.

Compounds	IC_{50} (ng/mL)	Cross-Reactivity (%)
DON	6.61	100
3-Ac-DON	142.1	4.7
15-Ac-DON	$>5 \times 10^3$	<0.2
DON-3-G	$>1 \times 10^4$	<0.1
NIV	$>1 \times 10^4$	<0.1
Fusarenon-X	$>1 \times 10^4$	<0.1
T-2 toxin	$>1 \times 10^4$	<0.1
ZEN	$>1 \times 10^4$	<0.1
AFB1	$>1 \times 10^4$	<0.1

3.3.4. Stability Determination

As shown in Figure 3, the values of B_0 (the value of absorbance without DON standard) and B/B_0 (%) (the ratio value of absorbance with 5 ng/mL DON and without DON standard) showed acceptable decreases during storage. The results showed that the kit had good stability and that its effective life was at least 12 months.

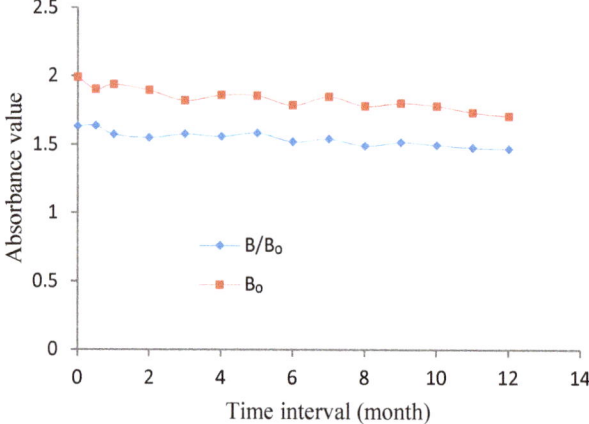

Figure 3. Stability of the dcELISA kit.

3.3.5. Matrix Effect Determination

As shown in Figure 4, the curves of the spiked samples of wheat bran, wheat flour, corn flour and corn were close to the DON standard curve by dilution of the extract solution multiple times, and their IC_{50} values were 8.81, 7.59, 6.22 and 5.7 ng/mL, respectively, indicating that the matrix interference was negligible. Therefore, the kit is functional for different substrates and can be used for detecting subsequent samples.

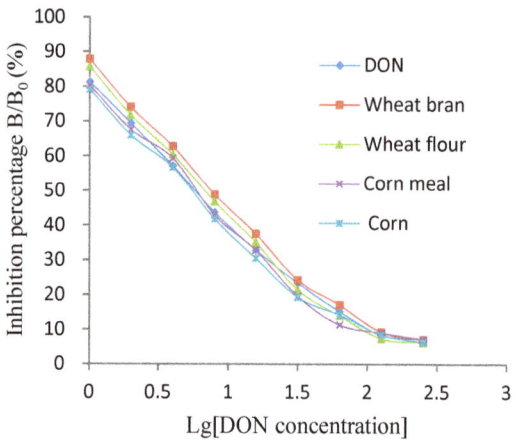

Figure 4. Effect of different samples' matrixes on the dcELISA kit.

3.4. Confirmation of the DON dcELISA Kit with HPLC

Table 5 shows that a total of 30 wheat samples from different provinces in China were tested using the assembled DON dcELISA kit and HPLC. The average value of detection with HPLC was in the range of 560.4 to 1049.1 ng/g, and the RSD ranged from 12.4% to 43.4% (the results of HPLC were corrected by a recovery of 85.7%). The average value of detection with the kit was in the range of 580.5 to 1020.3 ng/g, and the RSD ranged from 13% to 43.8%. The results showed that the test results of the kit were generally higher than those of HPLC. However, the test results of the kit in its linear range were in good agreement with those of HPLC.

Table 5. Comparison of screening results of 30 wheat samples detected by two different methods.

Province	dcELISA Kit			HPLC		
	Range (ng/g)	Average Value ± SD%	RSD (%)	Range (ng/g)	Average Value ± SD%	RSD (%)
Henan	309.5–1243.8	933.4 ± 324.1	34.7	277.1–1231.4	918.1 ± 304.7	33.2
Anhui	378.9–1230.7	810.5 ± 276.3	34.1	350.5–1218.4	796.6 ± 268.5	33.7
Hebei	696.4–1087.9	853 ± 119.4	14	681.9–1023.1	834.3 ± 103.1	12.4
Shandong	741.9–1258.4	1020.3 ± 168.2	16.5	752.8–1235.6	1049.1 ± 158.2	15.1
Jiangsu	591.2–891.2	713.4 ± 93	13	583.5–870.3	703.9 ± 92.4	13.1
Gansu	254.7–1080.5	580.5 ± 254.3	43.8	236.1–991.6	560.4 ± 243.3	43.4

Thirty samples of wheat were detected with the kit, 22 of which were found to contain DON, with a concentration range of 254.7 to 1258.4 ng/g. Four of the 30 samples were false suspect, with a false suspect rate of 13.3% (Table 6).

Table 6. Test results of wheat samples from different provinces by the kit and HPLC.

Province	Samples (Wheat)	dcELISA Kit	HPLC
Henan	1	+	+
	2	+	+
	3	-	-
	4	+	+
	5	+	-
Anhui	1	+	+
	2	-	+
	3	+	-
	4	+	+
	5	+	+
Hebei	1	+	+
	2	+	+
	3	-	-
	4	+	+
	5	-	+
Shandong	1	-	-
	2	+	+
	3	+	+
	4	+	+
	5	-	-
Jiangsu	1	+	+
	2	+	+
	3	-	-
	4	-	-
	5	+	+
Gansu	1	+	-
	2	+	+
	3	+	+
	4	+	+
	5	+	-

+, positive; -, negative; +, false negative; -, false suspect.

4. Discussions

4.1. Pretreatment of Biotoxin Samples

The pretreatment of samples increases the accuracy of HPLC and ELISA analyses. The samples were extracted and detected for DON; the pretreatment of samples for detection by ELISA was relatively simple, and direct filtration after extraction was sufficient for detection, while the pretreatment of samples for detection with HPLC required an immunoaffinity column. Yang et al. reported [37] that, the recovery rate of DON in ELISA was higher than 75%, and the RSD was 4.7% to 10.6% after the sample was filtered directly, while after passing through the immunoaffinity column, the recoveries of DON in HPLC and ELISA were the same when the spiked concentration of the standard was higher. It is concluded that the DON kit simplifies the process of sample pretreatment and purification. The results are accurate and reliable, and the detection steps are simple. It is very suitable for the rapid detection of a large number of samples. Moreover, ELISA detection technology has the advantages of limited interference, strong specificity, and short enzymatic reaction times, which shortens the whole detection time.

4.2. Determination of dcELISA Kit Performance

In this study, the dcELISA kit was assembled with an in-house-developed homemade high-affinity anti-DON monoclonal antibody. The kit performance metrics included sensitivity, accuracy, precision, specificity, stability and matrix effect, among others. Sensitivity determination can be calculated according to the method of Hayashi et al. [32]. The sensitivity of competitive ELISA is B/B_0 = 83.3%, which can also be calculated by the formula of limit of detection LOD (%) = [(X − 2SD)/X] × 100%. The method of B/B_0 = 83.3% was adopted in this experiment. The sensitivity was determined as 0.62 ng/mL, the detection limit was 1.0 ng/mL, and the detection range (IC_{10} to IC_{80}) was 1.0 to 113.24 ng/mL in the working buffer. According to the procedures of authentic sample pretreatment and extraction, the DON levels of samples were equivalent to a 100-fold dilution and the matrix effects were negligible. Thus, for the analysis of wheat samples, with a sensitivity of 62 ng/g, an LOD of 100 ng/g, and a detection range from 100 to 11,324 ng/g in authentic agricultural samples, the cross-reaction rate with 3-Ac-DON and 15-Ac-DON was 4.7%, less than 0.2%, respectively. The DON ELISA method established by the Ministry of Health in China has a detection limit of 5 ng/mL, and the detection range was 5 to 1000 ng/mL. It had been approved as the national recommended standard detection method of China. Therefore, the DON dcELISA kit assembled in our laboratory meets the domestic detection range and sensitivity standard requirements of DON analysis in food and feed. Compared with the commercial kits, the sensitivity and specificity is higher (3 ng/mL in the working buffer), the DON levels of samples were equivalent to a 100-fold dilution with an LOD of 300 ng/g in authentic agricultural samples, and the cross-reaction rate with 3-Ac-DON and 15-Ac-DON was less than 70%, less than 1%, respectively. Compared with the kit that was developed by Li et al. [12], it has higher sensitivity and specificity (4.9 ng/mL in the working buffer), the DON levels of samples were equivalent to a 40-fold dilution with an LOD of 200 ng/g in authentic agricultural samples, and the cross-reaction rate with 3-Ac-DON and 15-Ac-DON was 5.7%, less than 0.5%, respectively. Accuracy and precision are measured by spiked sample recovery (%) and relative standard deviation (RSD%). Generally, the recovery rate is between 70% and 140%. The average recovery was in the range of 77.1% to 107.0%, and the RSD was 4.2% to 11.9% in this experiment, which meets the national accuracy and precision test requirements, indicating that the kit could be used for the detection of actual samples. Therefore, by evaluating the recoveries and determining the DON content in wheat samples, it is proved that the developed dcELISA kit is accurate, reliable, and simple, and that it requires less instrumentation, and involves simple experimental steps for detecting DON content in food and feed. Compared with commercial kits, it is a more advanced detection method in China and abroad, providing a highly sensitive, economical and safe DON detection method.

4.3. Comparison of the Results of the dcELISA Kit and HPLC

The kit test results were generally higher than those of HPLC. The wheat samples from the farms in the six provinces of Henan, Anhui, Hebei, Shandong, Jiangsu and Gansu in China were analyzed for DON content using both the kit and HPLC. A total of 30 samples were randomly checked (five samples from each province). The average value of detection with HPLC was in the range of 560.4 to 1049.1 ng/g, and the RSD ranged from 12.4% to 43.4%. The average value of detection with the kit was in the range of 580.5 to 1020.3 ng/g, and the RSD ranged from 13% to 43.8%. Therefore, the results showed that the test results of the kit were generally higher than those of the HPLC. However, the test results of the kit in its linear range were in good agreement with those of HPLC. Therefore, the kit can be used for the determination of DON in food and feed. Antibodies are the basis of the ELISA kit detection method, which may lead to false positive or false negative results, while HPLC is commonly used as an accurate verification method. There were two main reasons why the kit test results were generally too high. First, high or low pH of the sample solution will affect the test results. Therefore, it is necessary to adjust the pH of the sample solution before detection. Some studies have found that when the pH of a sample extract is lower than 5, the structure of the enzyme in an enzyme-labeled antigen changes irreversibly, and most of its activity is lost, resulting in a reduced color reaction, which leads to false-positive results. The pH of the extracted solution was from 6 to 8 when the samples were purified in this experiment, which met the detection requirements of the kit. Therefore, the pH of the samples did not need to be adjusted when samples were detected. Second, the loss of sample in the pretreatment process of HPLC leads to low detection results (the results of HPLC in this experiment were corrected by a recovery of 85.7%).

5. Conclusions

In this experiment, a dcELISA kit method was established by using an anti-DON monoclonal antibody developed in our laboratory. The working concentrations of RaMIgG, anti-DON mAb and HRP-DON were optimized, and the performance of the developed kit was tested. Finally, a comparison of the results of the kit with those of HPLC shows that the developed kit has the same detection ability as HPLC. Therefore, the kit can be widely used for DON detection in food and feed.

Author Contributions: W.-J.Z. and Z.-L.W. designed the research and interpreted the results. L.H. and Y.-T.L. conducted the experiments and drafted the manuscript. Y.Z. and J.-M.L. collected and processed data. J.-Q.J., R.-F.L. and G.-Y.F. revised the manuscript. All authors have read and agreed to the published version of the manuscript.

Funding: This work was financially supported by the Modern Agricultural Science and Technology Tackling and Achievement Conversion Project of the Eighth Normal University [2018NY05]; the Twelfth Five-Year Plan of National Science and Technology Support Projects, "Research and Demonstration of Rapid Detection Technology for Hormone Drugs" [2014BAD13B05-01]; the Key Technological Research Projects of Henan Province in 2018 (Agriculture) [182102110222]; and the Program for Innovative Research Team (in Science and Technology) in University of Henan Province (20IRTSTHN025).

Conflicts of Interest: There are no conflict to declare.

References

1. Zuo, H.G.; Zhu, J.X.; Shi, L.; Zhan, C.R.; Guo, P.; Wang, Y.; Zhang, Y.M.; Liu, J.P. Development of a novel immunoaffinity column for the determination of deoxynivalenol and its acetylated derivatives in cereals. *Food Anal. Method.* **2018**, *11*, 2252–2260. [CrossRef]
2. Jiang, D.; Chen, J.; Li, F.; Li, W.; Wang, X. Deoxynivalenol and its acetyl derivatives in bread and biscuits in Shandong province in China. *Food Addit. Contam. B* **2017**, *11*, 43–48. [CrossRef] [PubMed]
3. Dezhao, K.; Xiaoling, W.; Yue, L.; Liqiang, L.; Shanshan, S.; Qiankun, Z.; Hua, K.; Chuanlai, X. Ultrasensitive and eco-friendly immunoassays based monoclonal antibody for detection of deoxynivalenol in cereal and feed samples. *Food Chem.* **2019**, *270*, 130–137. [CrossRef]
4. Ennouari, A.; Sanchis, V.; Marín, S.; Rahouti, M.; Zinedine, A. Occurrence of deoxynivalenol in durum wheat from morocco. *Food Control* **2013**, *32*, 115–118. [CrossRef]
5. Pestka, J.J.; Smolinski, A.T. Deoxynivalenol: Toxicology and potential effects on humans. *J. Toxicol. Environ. Heal B* **2005**, *8*, 39–69. [CrossRef]
6. Ji, F.; Li, H.; Xu, J.; Shi, J. Enzyme-linked immunosorbent-assay for deoxynivalenol (DON). *Toxins* **2011**, *3*, 968–978. [CrossRef]
7. Liao, Y.; Peng, Z.; Chen, L.; Nüssler, A.K.; Liu, L.; Yang, W. Deoxynivalenol, gut microbiota and immunotoxicity: A potential approach? *Food Chem. Toxicol.* **2018**, *112*, 342–354. [CrossRef]
8. Young, J.C.; Fulcher, R.G.; Hayhoe, J.H.; Scott, P.M.; Dexter, J.E. Effect of milling and baking on deoxynivalenol (vomitoxin) content of eastern canadian wheats. *J. Agric. Food Chem.* **1984**, *32*, 659–664. [CrossRef]

9. Tima, H.; Berkics, A.; Hannig, Z.; Ittzés, A.; Kecskésné Nagy, E.; Mohácsi-Farkas, C.; Kiskó, G. Deoxynivalenol in wheat, maize, wheat flour and pasta: Surveys in hungary in 2008–2015. *Food Addit. Contam. B* **2017**, *11*, 37–42. [CrossRef]
10. Arrúa Alvarenga, A.A.; Moura Mendes Arrua, J.; Cazal Martínez, C.C.; Arrúa Alvarenga, P.D.; Fernández Ríos, D.; Pérez Estigarribia, P.E.; Mohan Kohli, M. Deoxynivalenol screening in wheat-derived products in Gran Asunción, Paraguay. *J. Food Saf.* **2018**, *39*, 51–56. [CrossRef]
11. Iqbal, S.Z.; Alim, M.; Jinap, S.; Arino, A. Regulations for Food Toxins. In *Food Safety*; Springer International Publishing: Basel, Switzerland, 2016. [CrossRef]
12. Li, M.; Sun, M.; Hong, X.; Duan, J.; Du, D. Survey of Deoxynivalenol Contamination in Agricultural Products in the Chinese Market Using An ELISA Kit. *Toxins* **2019**, *11*, 6. [CrossRef] [PubMed]
13. China Food and Drug Administration. *China National Standard No. GB2761-2017*; Ministry of Health of P. R. China: Beijing, China, 2017.
14. Maragos, C.M.; Mccormick, S.P. Monoclonal antibodies for the mycotoxins deoxynivalenol and 3-acetyl-deoxynivalenol. *Food Agric. Immunol.* **2000**, *12*, 181–192. [CrossRef]
15. Wang, X.Y. Determination of deoxynivalenol in maize by using gas chromatography-electron capture detector. *J. Changzhi Univ.* **2006**, *23*, 7–10.
16. Righetti, L.; Galaverna, G.; Dall'Asta, C. Group detection of don and its modified forms by an elisa kit. *Food Addit. Contam. A* **2016**, *34*, 248–254. [CrossRef]
17. Wang, Z.P.; Wang, D.L.; Feng, Z.S.; Yang, J.; Wang, X.J.; Chen, X. Detection of deoxynivalenol in malting barley by immunoaffinity column clear up and high performance liquid chromatography. *Food Ferment. Ind.* **2008**, *34*, 137–139. [CrossRef]
18. Dos Santos, J.S.; Takabayashi, C.R.; Ono, E.Y.S.; Itano, E.N.; Mallmann, C.A.; Kawamura, O.; Hirooka, E.Y. Immunoassay based on monoclonal antibodies versus lc-ms: Deoxynivalenol in wheat and flour in southern brazil. *Food Addit. Contam. A* **2011**, *28*, 1083–1090. [CrossRef] [PubMed]
19. Zhang, R.S.; Zhou, Y.J.; Zhou, M.G. A sensitive chemiluminescence enzyme immunoassay for the determination of deoxynivalenol in wheat samples. *Anal. Methods UK* **2015**, *7*, 2196–2202. [CrossRef]
20. Li, C.L.; Wen, K. A universal multi-wavelength fluorescence polarization immunoassay for multiplexed detection of mycotoxins in maize. *Biosens. Bioelectron.* **2016**, *79*, 258–265. [CrossRef] [PubMed]
21. Zhang, J.; Gao, L.; Zhou, B.; Zhu, L.; Zhang, Y.; Huang, B. Simultaneous detection of deoxynivalenol and zearalenone by dual-label time-resolved fluorescence immunoassay. *J. Sci. Food Agric.* **2011**, *91*, 193–197. [CrossRef]
22. Yang, X.; Huang, Z.B.; He, Q.H.; Deng, S.Z.; Li, L.S.; Li, Y.P. Development of an immunochromatographic strip test for the rapid detection of deoxynivalenol in wheat and maize. *Food Chem.* **2010**, *119*, 834–839.
23. Burmistrova, N.A.; Rusanova, T.Y.; Yurasov, N.A.; Goryacheva, I.Y.; De Saeger, S. Multi-detection of mycotoxins by membrane based flow-through immunoassay. *Food Control* **2014**, *46*, 462–469. [CrossRef]
24. Kadota, T.; Takezawa, Y.; Hirano, S.; Tajima, O.; Maragos, C.M.; Nakajima, T.; Tanaka, T.; Kamata, Y.; Yoshiko, S.K. Rapid detection of nivalenol and deoxynivalenol in wheat using surface plasmon resonance immunoassay. *Anal. Chim. Acta* **2010**, *673*, 173–178. [CrossRef] [PubMed]
25. Yu, Q.; Li, H.; Li, C.L.; Zhang, S.X.; Shen, J.Z.; Wang, Z.H. Gold nanoparticles-based lateral flow immunoassay with silver staining for simultaneous detection of fumonisin B1 and deoxynivalenol. *Food Control* **2015**, *54*, 347–352. [CrossRef]
26. Qiu, Y.L.; He, Q.H.; Xu, Y.; Bhunia, A.K.; Tu, Z.; Chen, B.; Liu, Y.Y. Deoxynivalenol-mimic nanobody isolated from a naïve phage display nanobody library and its application in immunoassay. *Anal. Chim. Acta* **2015**, *887*, 201–208. [CrossRef] [PubMed]
27. Li, Y.; Shi, W.; Shen, J.; Zhang, S.; Cheng, L.; Wang, Z. Development of a rapid competitive indirect elisa procedure for the determination of deoxynivalenol in cereals. *Food Agric. Immunol.* **2012**, *23*, 41–49. [CrossRef]
28. Jia-Jia, T.; Xiao-Bing, L.I.; Guo-Wen, L.; Tao, K.; Dong-Na, L.I.; Liang, Z. Production of two anti-cadmium monoclonal antibodies by hock immunization. *Chin. Vet. Sci.* **2011**, *41*, 936–940.
29. Zeng, K.; Zou, Y.; Liu, J.; Wei, W.; Zhang, M.; Zhou, J.; Zhang, Z.; Gai, Z.K. Enzyme-linked immunosorbent assay for triclocarban in aquatic environments. *Water Sci. Technol. A J. Inter. Associ. Water Pollut. Res.* **2015**, *72*, 1682–1691. [CrossRef]
30. Kuang, H.; Xing, C.; Hao, C.; Liu, L.; Wang, L.; Xu, C. Rapid and highly sensitive detection of lead ions in drinking water based on a strip immunosensor. *Sensors* **2013**, *13*, 4214–4224. [CrossRef]

31. Xi, J.; Shi, Q. Development of an indirect competitive elisa kit for the detection of soybean allergenic protein gly m bd 28k. *Food Anal. Method* **2016**, *9*, 2998–3005. [CrossRef]
32. Hayashi, Y.; Matsuda, R.; Ito, K.; Nishimura, W.; Imai, K.; Maeda, M. Detection limit estimated from slope of calibration curve: An application to competitive elisa. *Anal. Sci.* **2005**, *21*, 167–169. [CrossRef]
33. Li, H. The Artificial Antigen Synthesis and Detection of Deoxynivalenol Using Indirect Competitive ELISA. Ph.D. Thesis, Nanjing Agricultural University, Nanjing, China, 2003.
34. Sakamoto, S.; Nagamitsu, R.; Yusakul, G.; Miyamoto, T.; Tanaka, H.; Morimoto, S. Ultrasensitive immunoassay for monocrotaline using monoclonal antibody produced by N, N' -carbonyldiimidazole mediated hapten-carrier protein conjugates. *Talanta* **2017**, *168*, 67–72. [CrossRef] [PubMed]
35. Niessen, L.; BoHm-Schrami, M.; Vogel, H.; Donhauser, S. Deoxynivalenol in commercial beer – screening for the toxin with an indirect competitive elisa. *Mycotoxin Res.* **1993**, *9*, 99–109. [CrossRef] [PubMed]
36. China Food and Drug Administration. *China National Standard No. GB5009.111-2016*; Ministry of Health of P. R. China: Beijing, China, 2016.
37. Yang, D.N.; Wang, H.F. ELISA and HPLC for detection of DON in wheat. *J. Anhui Agric. Sci.* **2013**, *41*, 1269–1270.

Sample Availability: Samples of the compounds are not available from the authors.

© 2019 by the authors. Licensee MDPI, Basel, Switzerland. This article is an open access article distributed under the terms and conditions of the Creative Commons Attribution (CC BY) license (http://creativecommons.org/licenses/by/4.0/).

Article

Enhancement of Solasodine Extracted from Fruits of *Solanum nigrum* L. by Microwave-Assisted Aqueous Two-Phase Extraction and Analysis by High-Performance Liquid Chromatography

Li Lin [1,†], Wen Yang [1,†], Xing Wei [1], Yi Wang [1], Li Zhang [1], Yunsong Zhang [1], Zhiming Zhang [2], Ying Zhao [1] and Maojun Zhao [1,*]

[1] College of Science, Sichuan Agricultural University, Yaan 625014, China; 14211@sicau.edu.cn (L.L.); Yangwen1611@163.com (W.Y.); weixing19960518@126.com (X.W.); 18728191338@163.com (Y.W.); zhangli@sicau.edu.cn (L.Z.); yaanyunsong@126.com (Y.Z.); zhaoying7767@163.com (Y.Z.)
[2] Maize research institute, Sichuan Agricultural University, Chengdu 611130, China; zhzhang@sicau.edu.cn
* Correspondence: Zhaomj_sicau@126.com; Tel.: +86-139-8161-7275
† These authors contributed equally to the paper.

Academic Editors: Clinio Locatelli, Marcello Locatelli and Dora Melucci
Received: 25 May 2019; Accepted: 11 June 2019; Published: 21 June 2019

Abstract: Background: Solasodine is a major bioactive ingredient in *Solanum nigrum* L. that has strong pharmacological characteristics. Therefore, the development of a simple and effective extraction method for obtaining solasodine is highly important. This study aims to provide a rapid and effective method for extracting solasodine from *Solanum nigrum* L. by microwave-assisted aqueous two-phase extraction (MAATPE). **Methods:** First, the high-performance liquid chromatography (HPLC) conditions were established for the detection of solasodine. Then, the aqueous two-phase system (ATPS) compositions were examined. On the basis of the results of single-factor experiments, for a better yield, response surface methodology (RSM) was used to optimize influential factors including the extraction temperature, extraction time and liquid-to-solid ratio. **Results:** The maximum extraction yield of 7.11 ± 0.08 mg/g was obtained at 44 °C, an extraction time of 15 min, and a liquid-to-solid ratio of 42:1 mL/g in the ATPS consisting of EtOH solvent, $(NH_4)_2SO_4$, and water (28:16:56, $w/w/w$). The extraction yield of the alkaloid obtained using this method was markedly higher than those of microwave-assisted extraction (MAE) and ultrasonic-assisted extraction (UAE). **Conclusions:** In this work, solasodine was extracted by MAATPE for the first time and a high yield was obtained. MAATPE is a simple, rapid, and green technique for extraction from medical plants. Thus, the present study will enable the development of a feasible extraction method of active alkaloids from *Solanum nigrum* L.

Keywords: microwave-assisted aqueous two-phase extraction; Solasodine; alkaloid; response surface methodology; high-performance liquid chromatography

1. Introduction

Solanum nigrum L. belongs to the Solanaceae family. As a typical traditional medicine in the Chinese pharmacopoeia, its leaves and fruits possess high medical value for treatments to clear heat, remove toxins, reduce swelling, and heal inflammation. Alkaloids are the major active constituents in this plant. In recent years, *Solanum nigrum* L. extracts, with antibacterial, antiviral, antioxidant, liver trauma treatment, and antineoplastic activities, have been extensively developed through a series of experiments, and considerable progress has been made [1–5]. Solasodine (in *Solanum nigrum* L.), the aglycone of steroidal alkaloids, can be obtained by hydrolysis and is a medically important component that is mainly used for antineoplastic drugs [6,7].

Solasodine is easily soluble in benzene, pyridine, and chloroform, soluble in ethanol, methanol, and acetone, and slightly soluble in water, and is closely connected to the cell wall in plants by aglycone. Conventional extraction methods, such as heat reflux extraction, Soxhlet extraction, and supercritical fluid extraction, have been developed to extract alkaloids from *Solanum nigrum* L. [8–11]. However, these processes suffer from obvious shortcomings, such as a high operating cost, poor extraction efficiency, long extraction time, and high usage of solvent in the products. Most recently, microwave-assisted extraction (MAE) has been widely applied to intensify the extraction and separation process [12] in the extraction of active substances from food, natural products, and traditional Chinese medicinal herbs. Microwave energy can enhance the penetration of the solvent into the matrix, expediting the release of bioactive compounds, which could significantly increase the extraction rate of the active ingredients from plants. The main advantages of MAE over the conventional extraction techniques are a short extraction time, smaller solvent consumption, and good selectivity [13–15].

An aqueous two-phase system (ATPS) can be spontaneously formed by mixing two water-soluble polymers or a water-soluble polymer and a salt aqueous solution at specific concentrations. Based on the differences in the compositions of the two phases, ATPS is divided into four types: polymer aqueous two-phase extraction [16], small molecule organic solvent aqueous two-phase extraction [17], surfactant aqueous two-phase extraction [18], and ionic liquid aqueous two-phase extraction [19]. Compared with other types of ATPS, the small molecular alcohol ATPS has outstanding advantages, such as a low cost, low viscosity, easy recovery, homogeneous phase extraction, and lack of phase emulsification. Therefore, aqueous two-phase extraction (ATPE) is considered to be a versatile and effective alternative to the conventional extraction methods. ATPE has been applied to the separation of compounds, such as the effective components of natural products [20–22], biological molecules [23–25], metal ions [26], and organic compounds [27]. Based on this technique, the novel microwave-assisted aqueous two-phase system extraction (MAATPE) method combines extraction and purification and offers some of the advantages of MAE and ATPE due to the demixing effect and microwave action, reduction of the extraction time, improved product yield, and increased purity of the extracted alkaloids [22,28]. Compared to ultrasound-assisted extraction (UAE) and other typical extraction methods, the MAATPE one-step extraction method is a promising and powerful alternative for the extraction and purification of alkaloids from *Solanum nigrum* L.

The application of MAATPE to the extraction of solasodine from *Solanum nigrum* L. has not been reported to date. The present study aimed to evaluate and maximize the potential and effectiveness of MAATPE as a rapid and effective method for the extraction of solasodine. After an initial screening, different ATPSs were tailored for the highest extraction efficiency, followed by the optimization of the extraction conditions carried out using the response surface methodology (RSM) to provide the most efficient process [29]. Moreover, to evaluate the feasibility and superiority of MAATPE, MAE and UAE methods, using an ethanol aqueous solution without salts, were investigated for comparison. Finally, high-performance liquid chromatography (HPLC) is known to be the most precise and sensitive detection method, and, therefore, HPLC was used to determine the efficiency of the solasodine extraction.

2. Results

2.1. Qualitative Analysis of Solasodine

As shown in Figure 1, the contents of the extracts from the top phase and bottom phase of the ATPS were determined by HPLC analysis and were compared with the pure standard of solasodine. Almost no alkaloids were present in the bottom phase. This is because alkaloids show good solubility in the organic solvents of the top phase [22].

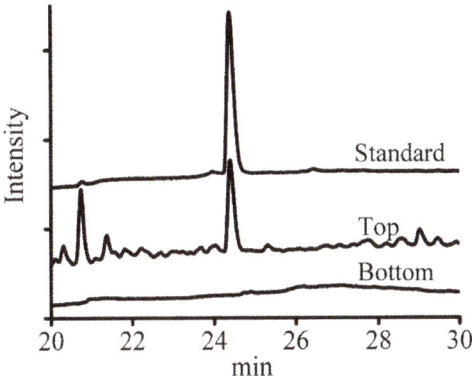

Figure 1. HPLC-UV chromatogram of solasodine from microwave-assisted aqueous two-phase system extraction (MAATPE).

2.2. Screening of the ATPS

The effects of different alcohols (PEG, EtOH, n-PrOH, and n-BuOH) and salts (($NH_4)_2SO_4$, K_2HPO_4, Na_2HPO_4, Na_2CO_3, and Na_2SO_4) on the solasodine extraction were studied. Among these, the bottom phase was easily saturated and precipitated for Na_2CO_3 and Na_2SO_4 due to the narrow range of the ATPS formed by these salts. The PEG/salt results showed that the PEG removal was complicated due to the high viscosity of PEG. Based on the preliminary experimental results, alcohols (EtOH, n-PrOH, and n-BuOH) and salts (($NH_4)_2SO_4$, K_2HPO_4, and Na_2HPO_4) were employed for the further study of the extraction performance. Here, nine types of ATPS were investigated, displaying diverse extraction abilities. As shown in Figure 2a, EtOH/($NH_4)_2SO_4$ was employed for the further optimization of the extraction conditions.

According to previous works [30,31], the composition of the ATPS lies within a small range of EtOH and ($NH_4)_2SO_4$ concentrations. Thus, the effect of the change in the EtOH content from 25% to 34% (w/w) was first studied under the conditions of an extraction temperature of 50 °C, extraction time of 10 min, and ($NH_4)_2SO_4$ concentration of 18% (w/w). The obtained results are presented in Figure 2b and show that the maximum extraction yield was obtained with an EtOH concentration of 28% (w/w). When 28% EtOH was used as the extraction solvent, the polarity of the solvent was close to the polarity of solasodine, resulting in a high yield of up to 5.14 mg/g. Interestingly, increasing the concentration of EtOH to 34% resulted in the partial solubility of ($NH_4)_2SO_4$ in the ATPS due to the interaction between the polarity and the volume ratio of the top phase, so that the ATPS cannot be formed at a constant temperature. Therefore, 28% EtOH was used in the subsequent experiments.

To optimize the yield, the ($NH_4)_2SO_4$ concentration was also investigated in the range of 14% to 20% (w/w), while the EtOH concentration was kept at 28% (w/w) and the other parameters were as described above. As shown in Figure 2c, the extraction yield increased from 5.44 mg/g to 5.83 mg/g as the ($NH_4)_2SO_4$ concentration changed from 14% to 16% and then decreased with the increasing ($NH_4)_2SO_4$ concentration for ($NH_4)_2SO_4$ concentrations greater than 18%. Therefore, 16% (w/w) ($NH_4)_2SO_4$ and 28% (w/w) EtOH were chosen as the optimal ATPS composition parameters.

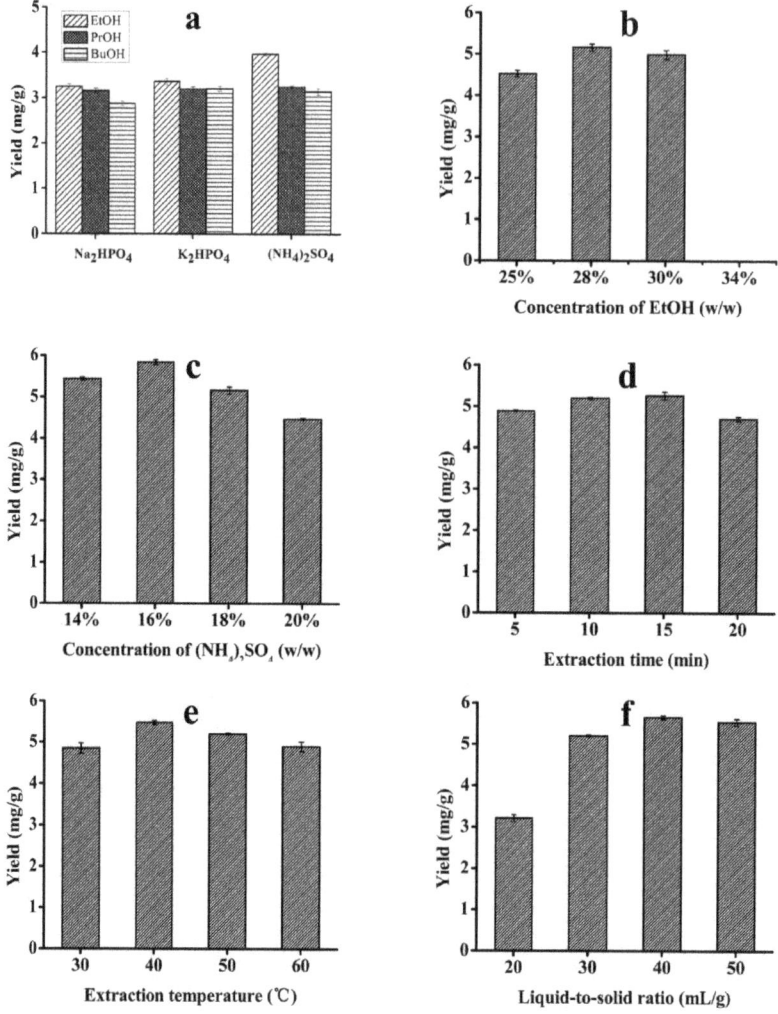

Figure 2. Different parameters of (**a**) type of aqueous two-phase system (ATPS), (**b**) EtOH concentration, (**c**) $(NH_4)_2SO_4$ concentration, (**d**) ultrasonication time, (**e**) extraction temperature, and (**f**) liquid-to-solid ratio for the extraction of solasodine using MAATPE.

2.3. Single Factor Experiment

2.3.1. Effect of Extraction Time

The effects of the extraction time on the extraction yield of solasodine were investigated for extraction times in the range of 5 to 20 min, while the liquid-to-solid ratio and temperature were kept at 30:1 and 50 °C, respectively. As shown in Figure 2d, increasing the time from 5 to 15 min had a positive effect on the yield, while a lower yield was obtained for an extraction time of 20 min. This may be because a sufficiently long time is necessary to release the active component from the herb cells, while too long a time leads to its degradation and decomposition. Thus, the extraction time of 15 min was used in the subsequent optimization of the extraction conditions.

2.3.2. Effect of Extraction Temperature

Figure 2e shows the effects of the extraction temperature on the yield of solasodine at temperatures ranging from 30 to 60 °C, with the other parameters selected according to the results described above. It was found that a relatively high yield was obtained, with the extraction efficiency decreasing gradually with higher temperatures for temperatures greater than 40 °C. It was concluded that higher temperatures led to a decrease in the surface tension and an increase in the vapour pressure within the microbubbles [32]. Thus, the greater penetration of the solvents and the solubility of solasodine improved the yield, but an excessive temperature led to the degradation and decomposition of the solasodine.

2.3.3. Effect of Liquid-to-Solid Ratio

The liquid-to-solid ratio is an important factor that can influence the extraction efficiency because the yield of solasodine is related to the contact area of the solid and liquid. As depicted in Figure 2f, the liquid-to-solid ratio was set to 20:1, 30:1, 40:1, and 50:1, respectively. The results obtained using these ratios indicate that the extraction efficiency was enhanced with the increasing liquid-to-solid ratio and reached a peak value at the liquid-to-solid ratio of 40:1; this is because an excess of solvent allows the largest contact area and makes the liquid penetrate into the solid more easily. However, when the amount of solvent reached a certain value, the solasodine was completely dissolved and the yield of solasodine increased only slightly with any further increase in the liquid-to-solid ratio. In contrast, the extraction equilibrium can be easily reached with insufficient solvent, possibly leading to incomplete penetration and poor extraction yield [33]. Therefore, the liquid-to-solid ratio of 40:1 was chosen for the subsequent experiments.

2.4. Optimization of the Procedure by RSM

Based on the results of the single-factor experiments, three parameters, namely, temperature, time, and liquid-to-solid ratio, were chosen for a further study of the interactions between the various factors by the RSM approach based on the Box-Behnken design (BBD), which is a collection of mathematical and statistical techniques. Thus, 17 experiments (runs 1–17, 12 factorial points, and 5 central points) were run, and the experimental values obtained for the yield of solasodine are presented in Table 1. The response variables were fitted to a second-order polynomial model equation estimated by the RSM:

$$Y = \beta_0 + \sum_{i=1}^{3} \beta_i X_i + \sum_{i=1}^{3} \beta_{ii} X_i^2 + \sum_{i=1}^{3} \sum_{j=i+1}^{3} \beta_{ij} X_i X_j, \tag{1}$$

where Y is the response variable, which is the yield of solasodine, X_i and X_j are the independent variables affecting the response, and β_0, β_i, β_{ii}, β_{ij} are the regression coefficients of the intercept, linear, quadratic, and interaction terms, respectively.

The significance of each coefficient was determined in the regression model. An analysis of variance for the evaluation of the second-order model is presented in Table 2. The model p-value was used to evaluate the significance of each coefficient, which is necessary to understand the pattern of the interactions between the independent variables. The p-value of solasodine was much smaller than 0.0001, indicating that the corresponding coefficients were highly significant. The R^2 parameter was 0.9675, showing that more than 96.75% of the experimental data were fitted by the model. These results reveal that the regression model is highly reliable.

Table 1. The experimental results of the Box-Behnken Design: X_1, X_2 and X_3 are temperature, time, and liquid-to-solid ratio, respectively.

Run	Factors			Yield (mg/g)
	X_1 (°C)	X_2 (min)	X_3 (mL/g)	
1	30	20	40	4.76
2	50	15	30	4.72
3	30	15	30	4.65
4	50	10	40	5.28
5	40	10	30	4.21
6	50	20	40	5.55
7	50	15	50	6.50
8	40	15	40	6.92
9	40	15	40	6.92
10	40	20	50	4.75
11	40	10	50	4.55
12	40	15	40	6.92
13	40	15	40	6.92
14	40	20	30	4.94
15	30	15	50	4.35
16	30	10	40	4.40
17	40	15	40	6.92

Table 2. Analysis of variance (ANOVA) for the second-order response surface model.

Source	Sum of Squares	Degrees of Freedom	Mean Square	F-Value	p-Value	Remarks
Model	18.71	9	2.08	53.87	<0.0001	significant
A(temperature)	1.89	1	1.89	49.01	0.0002	significant
B(time)	0.3	1	0.3	7.88	0.262	
C(liquid-to-solid ratio)	0.33	1	0.33	8.6	0.219	
AB	0.002	1	0.002	0.052	0.8254	
AC	1.08	1	1.08	28.02	0.0011	significant
BC	0.07	1	0.07	1.82	0.2194	
A2	2.31	1	2.31	59.74	0.0001	significant
B2	5.89	1	5.89	152.54	<0.0001	significant
C2	5.33	1	5.33	138.07	<0.0001	significant
Residual	0.27	7	0.039			
Lack of fit	0.27	3	0.09			
Pure error	0	4	0			
Cor total	18.98	16				

Based on the quadratic model, the optimal values of the independent variables and the response variable for the solasodine extraction were calculated as follows: an extraction temperature (X_1) of 44 °C, extraction time (X_2) of 15 min, and liquid to-solid ratio (X_2) of 42:1 and a maximum predicted value of 7.04 mg/g for the yield. Under these optimal extraction conditions, triplicate validation experiments were carried out, and the average obtained extraction yield of solasodine was 7.11 ± 0.08 mg/g. This value agrees fairly well with the predicted result, indicating that the proposed method can be applied to optimize the conditions of solasodine extraction.

2.5. Comparison of MAATPE with UAE and MAE

To further evaluate the extraction efficiency of MAATPE, two conventional methods, MAE and UAE, were employed for comparison, following the procedures described in previous reports [34,35], and the results are summarized in Table 3. The extraction yield of solasodine by MAATPE is dramatically higher than those for UAE and MAE, which were performed using aqueous ethanol as the extractant without salts with an extraction time of 30 min. The use of ATPS by adding $(NH_4)_2SO_4$ to the ethanol-water significantly improved the extraction efficiency. Thus, the proposed MAATPE method is an effective approach for the extraction of solasodine.

Table 3. Comparison of MAATPE with conventional methods with regard to the extraction yield of solasodine.

Method	Solvent	Time (min)	Tem. (°C)	Liquid-to-Solid Ratio (mL/g)	Yield (mg/g)
UAE	EtOH/water	30	25	30:1	3.39
MAE	EtOH/water	60	30	20:1	3.36
MAATPE	ATPS	15	44	44:1	7.11

3. Discussion

3.1. Effect of ATPS Composition and Concentration

The ATPS consisting of inorganic salts and small molecule alcohols are formed due to the competition between the salts and alcohols for the water molecules to form their associated hydrates [36]. Therefore, the solubility of the salts in water and the molecular weight of the alcohols determined the partition coefficients of the two phases in the ATPS. As shown in Figure 2a, the maximum extraction yield was obtained using EtOH/$(NH_4)_2SO_4$ as the extracting agent, owing to its good layering effect, high solubility for the alkaloid components in the plant materials, and high stability [30]. For the same alcohol, $(NH_4)_2SO_4$ has higher solubility in water and a stronger water molecule competitiveness, increasing the alcohol phase concentration and improving the yield of the target component. Similarly, the ATPS formed from EtOH achieved high yields in the three alcohols, which can be explained by the principle of similar compatibility. As for the ATPS concentration, based on the similarity and intermiscibility theory, the solute derived from the plants is easily dissolved when the polarities of the solvent and solute are similar [30]. These results also demonstrated that salts could improve the conductivity and microwave action in the MAATPE, while the volume of the top phase decreased with the increase in the $(NH_4)_2SO_4$ concentration, leading to reductions in the amounts of the target constituents [22,30].

3.2. Response Surface Analysis

For a better understanding of the effects of each of the extraction factors and the optimal conditions for obtaining the maximum extraction yield, three three-dimensional profiles were plotted for analysing the interactions of the various process factors using Design-Expert 8.0 software, as shown in Figure 3. The three-dimensional profiles show how the three pairs of extraction parameters affect the extraction yield of solasodine. Figure 3a–c presents the combined effects of temperature (X_1) and time (X_2), temperature (X_1) and liquid-to-solid ratio (X_3), and time (X_2) and liquid-to-solid ratio (X_3), respectively. All three surfaces are top-convex with a maximum point in the centre of the experimental domain, indicating that the ranges of the factors were chosen correctly.

Figure 3a shows that the solasodine yield improved with increases in the temperature and time, as indicated by the positive coefficients (Table 4), while the negative interaction between X_1 and X_2 produced maximum values at temperatures higher than 40 °C and times longer than 15 min. According to the results presented in Figure 3b, similar conclusions can be reached based on the positive coefficients of X_1, X_3, and X_1X_3; therefore, a liquid-to-solid ratio higher than 30:1 is a better choice in the optimization. Figure 3c also shows the presence of significant interactions between X_2 and X_3.

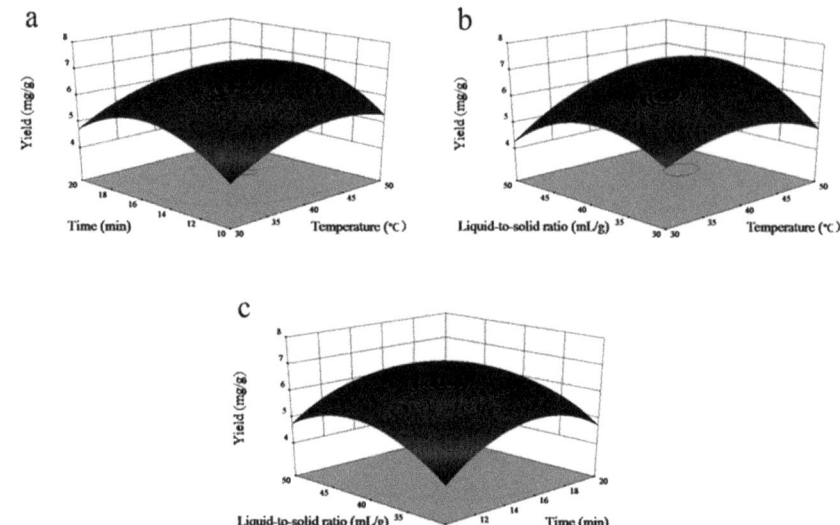

Figure 3. Response surface curves showing the effects of (**a**) time and temperature, (**b**) liquid-to-solid ratio and temperature, and (**c**) liquid-to-solid ratio and time on the extraction efficiency.

Table 4. Estimated regression coefficients of the fitted second-order polynomial equation.

Factors	Coefficient	Df	Standard Error	95% Low	95% High	CL VIF
intercept	6.92	11	0.088	6.71	7.13	1
A- temp.	0.49	1	0.069	0.32	0.65	1
B- time	0.2	1	0.069	0.031	0.36	1
C- liquid-to-solid ratio	0.2	1	0.069	0.04	0.37	1
AB	−0.023	1	0.098	−0.25	0.21	1
AC	0.52	1	0.098	0.29	0.75	1
BC	−0.13	1	0.098	−0.36	0.1	1
A2	−0.74	1	0.096	−0.97	−0.51	1
B2	−1.18	1	0.096	−1.41	−0.96	1.01
C2	−1.12	1	0.096	−1.35	−0.9	1.01

4. Experimental

The minimum standards of reporting checklist contains details of the experimental design, statistics, and resources used in this study.

4.1. Chemicals and Plant Material

The dried herb fruit of *Solanum nigrum* L. was purchased from Baicaofang Co., Ltd. (Hebei, China). The samples were powdered, sieved (40 mesh), and placed in a desiccator at room temperature. Solasodine (purity ≥ 97%) was purchased from Chengdu Must Bio-Technology Co., Ltd. (Chengdu, China). HPLC grade acetonitrile was purchased from Thermo Fisher Scientific (Shanghai, China) and used in the HPLC analysis. Ammonium sulphate, anhydrous ethanol, and other chemicals of analytical grade were purchased from Chengdu Kelon Science Co., Ltd. (Chengdu, China).

4.2. Instruments and Analytical Methods

MAATPE experiments were carried out using a microwave extraction system (MAS-I) purchased from Sineo Microwave Chemistry Technology Co., Ltd. (Shanghai, China). The HPLC (LC-20A) was purchased from Shimadzu Corp. (Kyoto, Japan). The yield of solasodine in the extracts was determined using an Agilent C18 Column (250 mm × 4.6 mm, 5 μm, Santa Clara, CA, USA) at 40 °C with a UV detector at 210 nm. The injection volume was 10 μL, and gradient elution was performed at a flow rate of 1 mL/min over 50 min. The mobile phase, composed of solution A (0.1% phosphoric acid) and solution B (acetonitrile), was delivered as follows: 0 min, 10% (B); 30 min, 60% (B); 35 min, 90% (B); 40 min, 90% (B); 45 min, 10% (B); and 50 min, stop.

4.3. Extraction Procedure

4.3.1. Preparation of the Aqueous Two-Phase System

According to a method reported in a previous study [21], different solvents and salts were employed for the formation of ATPS under particular mixing ratios. An appropriate amount of the inorganic salt was dissolved in deionized water and was then mixed with a certain volume of alcohol using a vortex stirrer. The extraction agent was obtained when the mixture showed two-phase separation.

4.3.2. MAATPE Procedure

The optimal extraction conditions are described as an example: the herb (0.71 g) powders and ATPS (30 mL) solvents (EtOH/$(NH_4)_2SO_4$, 28:16, w/w) were added to a flask that was then placed in a microwave extraction system equipped with a cooling tube. The extraction was performed at 44 °C for 15 min, and the microwave power was set at 500 W in all of the related experiments, unless otherwise indicated. After cooling to room temperature, the extract solution was filtered to remove the herb residues, and the filtrate was held in a separatory funnel to allow phase separation. Then, the top phase was separated and concentrated to obtain the residues that were used in the subsequent hydrolytic reaction without any purification.

4.3.3. Hydrolytic Procedure

The glucoside bond is generally due to aldolization, and in the hydrolysis mechanism, the chemical bonds between aglycone and glycosyl react easily and are broken under acidic conditions [7], as shown in Figure 4. Thus, the hydrolytic procedures were carried out as follows: the concentration derived from MAATPE was mixed with an HCl solution (20 mL, 2 mol/L) and anhydrous ethanol (20 mL). The hydrolysis temperature was kept at 100 °C for 2 h, and then the mixture was cooled. The acidity of the mixture was adjusted to pH = 10~11 to obtain free solasodine [37]. The solvent was removed by the vacuum-rotary evaporation procedure to obtain the residues. Subsequently, desalination was performed by adding methanol (10 mL) and centrifuging at room temperature for 1 min. Then, the supernatant (5 mL) was removed and diluted with methanol to 25 mL for detection. For reproducibility, the results reported for the solasodine extraction efficiency were the averages of three repeated trials ($n = 3$).

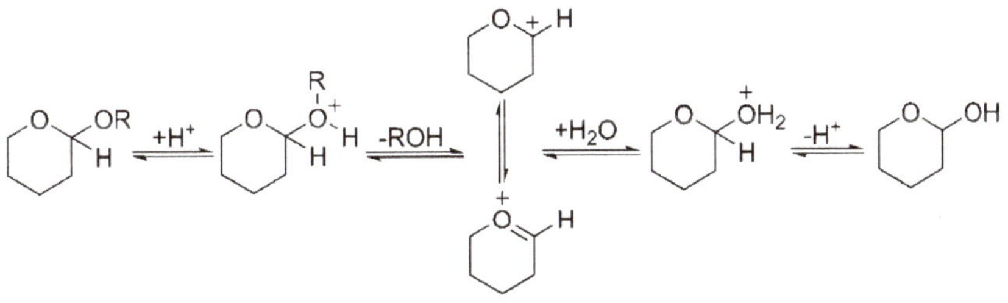

Figure 4. Hydrolysis mechanism of steroidal alkaloids, where R represents glycosyl.

4.4. Optimization of Extraction Conditions

The types of ATPS were screened in our initial experiments, together with four other main variables that affect the extraction efficiency, namely, the ATPS solvent concentration, extraction time, temperature, and liquid-to-solid ratio. Thus, the extraction conditions of MAATPE were first improved using single-factor experiments because the concentration of the solvent in the ATPS is small and was difficult to control in the RSM. Then, the RSM was applied to optimize the experimental conditions, including the extraction temperature, extraction time, and liquid-to-solid ratio. Design-Expert 8.0.6 software (Stat-Ease, Inc., Minneapolis, MN, USA) was employed not only for the experimental design but also for statistical analysis and regression modelling. The details of the RSM procedure are provided in Table 1. The parameters were screened based on the results of the single-factor experiments.

4.5. Conventional Extraction Methods

To compare the extraction efficiency of MAATPE to those of the conventional techniques, UAE and MAE were carried out at the optimized conditions determined in previous studies reported in the literature [34,35]. The UAE procedure at the optimal conditions was as follows: the sample powders (1.0 g) were mixed with aqueous ethanol (30 mL, 9:1, v/v) and extracted for 30 min using an ultrasonic generator (SB-600DTD, Scientz Biotechnology Co., Ltd., Ningbo, China). The MAE experiments were performed using the microwave device described above. The sample powders (1.0 g) were added to the ethanol aqueous solution (20 mL, 3:2, v/v) and extracted for 60 min. After extraction, the mixture was filtrated and the top phase was removed for analysis. The hydrolytic procedure was the same as that described in Section 2.3.3.

4.6. Statistical Analysis

All the experiments were carried out in triplicate. Design-Expert 8.0.6 software was employed not only for the experimental design, but also for statistical analysis and regression modelling. The Student's t-test permitted the checking of the statistical significance of the regression coefficient, and Fisher's F-test determined the second-order model equation at a probability (p) of 0.001, 0.01 or 0.05.

5. Conclusions

In the present study, MMAATPE was firstly introduced for improving the extraction yield of Solasodine from fruits of *Solanum nigrum* L., integrating MAE with ATPE into a one-step procedure and provided a rapid effective method for extracting Solasodine. Several factors were evaluated for the selection of the suitable ATPS and extraction methods by means of single factor experiment and RSM. The results indicated that the ATPS consisting of EtOH solvent, $(NH_4)_2SO_4$ and water (28:16:56, w/w/w) provided superior extraction yields with less time consumption compared with the traditional method. Thus, this approach will enable the development of a feasible extraction method of alkaloid activities from fruits of *Solanum nigrum* L. In summary, MAATPE exhibits a simple, rapid, and green technique for the extraction in medical plants.

Author Contributions: L.L. analyzed the data, and wrote the article. W.Y. performed the experiments and wrote the article. X.W. and Y.W. performed the experiments. L.Z. and Y.Z. (Yunsong Zhang) did the chemometric analysis. Y.Z. (Ying Zhao) did sample detection. Z.Z. collected the plant material. L.L. and M.Z. conceived and designed the experiments. All authors read and approved the final manuscript.

Funding: This research was funded by the Sichuan Provincial Science and Technology Department Key Program, grant number 2018JZ0028.

Acknowledgments: The authors thank the anonymous reviewers for their insightful comments and careful corrections.

Conflicts of Interest: The authors declare no conflict of interest.

References

1. Shen, K.H.; Hung, J.H.; Chang, C.W.; Weng, Y.T.; Wu, M.J.; Chen, P.S. Solasodine inhibits invasion of human lung cancer cell through downregulation of miR-21 and MMPs expression. *Chem. Boil. Interact.* **2017**, *268*, 129–135. [CrossRef] [PubMed]
2. Muthuvel, A.; Adavallan, K.; Balamurugan, K.; Krishnakumar, N. Biosynthesis of gold nanoparticles using *Solanum nigrum* leaf extract and screening their free radical scavenging and antibacterial properties. *Biomed. Prev. Nutr.* **2014**, *4*, 325–332. [CrossRef]
3. Tariq Javed, U.A.A.; Sana Riaz, S.R.; Sheikh, R. In-vitro antiviral activity of *Solanum nigrum* against Hepatitis C Virus. *Virol. J.* **2011**, *8*, 8–26.
4. Jiang, Q.-Y.; Tan, S.-Y.; Zhuo, F.; Yang, D.-J.; Ye, Z.-H.; Jing, Y.-X. Effect of Funneliformis mosseae on the growth, cadmium accumulation and antioxidant activities of *Solanum nigrum*. *Appl. Soil Ecol.* **2016**, *98*, 112–120. [CrossRef]
5. Kibichiy, S.E.; Jane, M.; Raymond, M.; Scolastica, K.; James, K.; John, K.; Raphael, N. Effects of crude extracts of *Solanum nigrum* on the Liver pathology and Survival time in Trypanosoma brucei rhodesiense infected mice. *Sci. J. Microbiol.* **2013**, 242–248. [CrossRef]
6. Patel, K.; Singh, R.B.; Patel, D.K. Medicinal significance, pharmacological activities, and analytical aspects of solasodine: A concise report of current scientific literature. *J. Acute Dis.* **2013**, *2*, 92–98. [CrossRef]
7. Dou, M.; He, X.-H.; Sun, Y.; Peng, F.; Liu, J.-Y.; Hao, L.-L.; Yang, S.-L. Controlled acid hydrolysis and kinetics of flavone C-glycosides from trollflowers. *Chin. Chem. Lett.* **2015**, *26*, 255–258. [CrossRef]
8. Fulzele, D.P.; Satdive, R.K. Comparison of techniques for the extraction of the anti-cancer drug camptothecin from *Nothapodytes foetida*. *J. Chromatogr. A* **2005**, *1063*, 9–13. [CrossRef]
9. Cassel, E.; Vargas, R.M.F.; Brun, G.W.; Almeida, D.E.; Cogoi, L.; Ferraro, G.; Filip, R. Supercritical fluid extraction of alkaloids from *Ilex paraguariensis* St. Hil. *J. Food Eng.* **2010**, *100*, 656–661. [CrossRef]
10. Yang, L.; Wang, H.; Zu, Y.-G.; Zhao, C.; Zhang, L.; Chen, X.; Zhang, Z. Ultrasound-assisted extraction of the three terpenoid indole alkaloids vindoline, catharanthine and vinblastine from *Catharanthus roseus* using ionic liquid aqueous solutions. *Chem. Eng. J.* **2011**, *172*, 705–712. [CrossRef]
11. Klein-Júnior, L.C.; Vander Heyden, Y.; Henriques, A.T. Enlarging the bottleneck in the analysis of alkaloids: A review on sample preparation in herbal matrices. *TrAC Trends Anal. Chem.* **2016**, *80*, 66–82. [CrossRef]
12. Ma, F.-Y.; Gu, C.-B.; Li, C.-Y.; Luo, M.; Wang, W.; Zu, Y.-G.; Li, J.; Fu, Y.-J. Microwave-assisted aqueous two-phase extraction of isoflavonoids from *Dalbergia odorifera* T. Chen leaves. *Sep. Purif. Technol.* **2013**, *115*, 136–144. [CrossRef]
13. Wu, Z.; Ruan, H.; Wang, Y.; Chen, Z.; Cui, Y. Optimization of microwave-assisted extraction of puerarin from radix puerariae using response surface methodology. *Sep. Sci. Technol.* **2013**, *48*, 1657–1664. [CrossRef]
14. Fliniaux, O.; Corbin, C.; Ramsay, A.; Renouard, S.; Beejmohun, V.; Doussot, J.; Falguieres, A.; Ferroud, C.; Lamblin, F.; Laine, E.; et al. Microwave-assisted extraction of herbacetin diglucoside from flax (*Linum usitatissimum* L.) seed cakes and its quantification using an RP-HPLC-UV system. *Molecules* **2014**, *19*, 3025–3037. [CrossRef] [PubMed]
15. Nkhili, E.; Tomao, V.; El Hajji, H.; El Boustani, E.S.; Chemat, F.; Dangles, O. Microwave-assisted water extraction of green tea polyphenols. *Phytochem. Anal. PCA* **2009**, *20*, 408–415. [CrossRef] [PubMed]
16. Madeira, P.P.; Reis, C.A.; Rodrigues, A.E.; Mikheeva, L.M.; Chait, A.; Zaslavsky, B.Y. Solvent properties governing protein partitioning in polymer/polymer aqueous two-phase systems. *J. Chromatogr. A* **2011**, *1218*, 1379–1384. [CrossRef] [PubMed]
17. Wu, Y.-T.; Pereira, M.; Venâncio, A.; Teixeira, J. Recovery of endo-polygalacturonase using polyethylene glycol-salt aqueous two-phase extractionwith polymer recycling. *Bioseparation* **2000**, *9*, 247–254. [CrossRef]
18. Minuth, T.; Thömmes, J.; Kula, M.R. Extraction of cholesterol oxidase from Nocardia rhodochrous using a nonionic surfactant-based aqueous two-phase system. *J. Biotechnol.* **1995**, *38*, 151–164. [CrossRef]
19. Soto, A.; Arce, A.; Khoshkbarchi, M. Partitioning of antibiotics in a two-liquid phase system formed by water and a room temperature ionic liquid. *Sep. Purif. Technol.* **2005**, *44*, 242–246. [CrossRef]
20. Wu, X.; Liang, L.; Zou, Y.; Zhao, T.; Zhao, J.; Li, F.; Yang, L. Aqueous two-phase extraction, identification and antioxidant activity of anthocyanins from mulberry (*Morus atropurpurea* Roxb.). *Food Chem.* **2011**, *129*, 443–453. [CrossRef]

21. Du, L.-P.; Cheong, K.-L.; Liu, Y. Optimization of an aqueous two-phase extraction method for the selective separation of sulfated polysaccharides from a crude natural mixture. *Sep. Purif. Technol.* **2018**, *202*, 290–298. [CrossRef]
22. Zhang, W.; Liu, X.; Fan, H.; Zhu, D.; Wu, X.; Huang, X.; Tang, J. Separation and purification of alkaloids from *Sophora flavescens* Ait. by focused microwave-assisted aqueous two-phase extraction coupled with reversed micellar extraction. *Ind. Crop. Prod.* **2016**, *86*, 231–238. [CrossRef]
23. Amid, M.; Shuhaimi, M.; Islam Sarker, M.Z.; Abdul Manap, M.Y. Purification of serine protease from mango (*Mangifera Indica Cv. Chokanan*) peel using an alcohol/salt aqueous two phase system. *Food Chem.* **2012**, *132*, 1382–1386. [CrossRef] [PubMed]
24. Rosa, P.A.; Azevedo, A.M.; Sommerfeld, S.; Backer, W.; Aires-Barros, M.R. Aqueous two-phase extraction as a platform in the biomanufacturing industry: Economical and environmental sustainability. *Biotechnol. Adv.* **2011**, *29*, 559–567. [CrossRef] [PubMed]
25. Ruiz-Ruiz, F.; Benavides, J.; Aguilar, O.; Rito-Palomares, M. Aqueous two-phase affinity partitioning systems: Current applications and trends. *J. Chromatogr. A* **2012**, *1244*, 1–13. [CrossRef] [PubMed]
26. Liu, X.; Gao, Y.; Tang, R.; Wang, W. On the extraction and separation of iodide complex of cadmium(II) in propyl-alcohol ammonium sulfate aqueous biphasic system. *Sep. Purif. Technol.* **2006**, *50*, 263–266. [CrossRef]
27. Jiang, B.; Li, Z.-G.; Dai, J.-Y.; Zhang, D.-J.; Xiu, Z.-L. Aqueous two-phase extraction of 2,3-butanediol from fermentation broths using an ethanol/phosphate system. *Process Biochem.* **2009**, *44*, 112–117. [CrossRef]
28. Chen, Z.; Zhang, W.; Tang, X.; Fan, H.; Xie, X.; Wan, Q.; Wu, X.; Tang, J.Z. Extraction and characterization of polysaccharides from Semen Cassiae by microwave-assisted aqueous two-phase extraction coupled with spectroscopy and HPLC. *Carbohydr. Polym.* **2016**, *144*, 263–270. [CrossRef]
29. Povilaitis, D.; Venskutonis, P.R. Optimization of supercritical carbon dioxide extraction of rye bran using response surface methodology and evaluation of extract properties. *J. Supercrit. Fluids* **2015**, *100*, 194–200. [CrossRef]
30. Zhang, W.; Zhu, D.; Fan, H.; Liu, X.; Wan, Q.; Wu, X.; Liu, P.; Tang, J.Z. Simultaneous extraction and purification of alkaloids from *Sophora flavescens* Ait. by microwave-assisted aqueous two-phase extraction with ethanol/ammonia sulfate system. *Sep. Purif. Technol.* **2015**, *141*, 113–123. [CrossRef]
31. Reschke, T.; Brandenbusch, C.; Sadowski, G. Modeling aqueous two-phase systems: III. Polymers and organic salts as ATPS former. *Fluid Phase Equilibria* **2015**, *387*, 178–189. [CrossRef]
32. Zhao, B.; Zhang, J.; Guo, X.; Wang, J. Microwave-assisted extraction, chemical characterization of polysaccharides from *Lilium davidii* var. *unicolor* Salisb and its antioxidant activities evaluation. *Food Hydrocoll.* **2013**, *31*, 346–356.
33. Yang, Z.; Tan, Z.; Li, F.; Li, X. An effective method for the extraction and purification of chlorogenic acid from ramie (*Boehmeria nivea* L.) leaves using acidic ionic liquids. *Ind. Crop. Prod.* **2016**, *89*, 78–86. [CrossRef]
34. Me, H.-W.; Xie, C.-Y.; Wu, H.-J.; Feng, L.; Zhao, F.-C.; Zhang, X.-Y. Study on extraction of alkaloids from *Solanum nigrum* L. fruits. *For. By-Prod. Spec. China* **2013**, *1*, 14–18.
35. Luan, G.Y.; Yang, Y.H. Study on ultrasonic-assisted extraction of total steroidal alkaloid from *Solanum nigrum* L. *J. Jilin Inst. Chem. Technol.* **2015**, *10*, 1158–1169.
36. Soares, R.R.; Azevedo, A.M.; Van Alstine, J.M.; Aires-Barros, M.R. Partitioning in aqueous two-phase systems: Analysis of strengths, weaknesses, opportunities and threats. *Biotechnol. J.* **2015**, *10*, 1158–1169. [CrossRef] [PubMed]
37. Jiang, X.Y.; Yang, H.; Zhao, Y. The Determination of Steroidal Alkaloid Content in *Solanum nigrum* L. *Food Sci.* **2006**, *27*, 224–227.

Sample Availability: Not available.

© 2019 by the authors. Licensee MDPI, Basel, Switzerland. This article is an open access article distributed under the terms and conditions of the Creative Commons Attribution (CC BY) license (http://creativecommons.org/licenses/by/4.0/).

Article

Development of a Method for the Quantification of Clotrimazole and Itraconazole and Study of Their Stability in a New Microemulsion for the Treatment of Sporotrichosis

Patricia Garcia Ferreira [1], Carolina Guimarães de Souza Lima [2], Letícia Lorena Noronha [1], Marcela Cristina de Moraes [2], Fernando de Carvalho da Silva [2], Alessandra Lifsitch Viçosa [3], Débora Omena Futuro [1] and Vitor Francisco Ferreira [1,*]

[1] Departamento de Tecnologia Farmacêutica, Faculdade de Farmácia, Universidade Federal Fluminense, Niterói-RJ 24241-000, Brazil; patricia.pharma@yahoo.com.br (P.G.F.); leticianoronha95@gmail.com (L.L.N.); dfuturo@id.uff.br (D.O.F.)
[2] Departamento de Química Orgânica, Instituto de Química, Universidade Federal Fluminense, Niterói-RJ 24210-141, Brazil; carolgslima@gmail.com (C.G.d.S.L.); mcmoraes@id.uff.br (M.C.d.M.); gqofernando@vm.uff.br (F.d.C.d.S.)
[3] Fundação Oswaldo Cruz (FIOCRUZ), Farmanguinhos-Manguinhos, Avenida Sinzenando Nabuco 100, Rio de Janeiro-RJ 21045-900, Brazil; alessandra.vicosa@far.fiocruz.br
* Correspondence: vitorferreira@id.uff.br; Tel.: +55-21-998578148

Academic Editors: Clinio Locatelli, Marcello Locatelli and Dora Melucci
Received: 6 June 2019; Accepted: 20 June 2019; Published: 25 June 2019

Abstract: Sporotrichosis occurs worldwide and is caused by the fungus *Sporothrix brasiliensis*. This agent has a high zoonotic potential and is transmitted mainly by bites and scratches from infected felines. A new association between the drugs clotrimazole and itraconazole is shown to be effective against *S. brasiliensis* yeasts. This association was formulated as a microemulsion containing benzyl alcohol as oil, Tween® 60 and propylene glycol as surfactant and cosurfactant, respectively, and water. Initially, the compatibility between clotrimazole and itraconazole was studied using differential scanning calorimetry (DSC), thermogravimetric analysis (TG), Fourier transform infrared spectroscopy (FTIR), and X-ray powder diffraction (PXRD). Additionally, a simple and efficient analytical HPLC method was developed to simultaneously determine the concentration of clotrimazole and itraconazole in the novel microemulsion. The developed method proved to be efficient, robust, and reproducible for both components of the microemulsion. We also performed an accelerated stability study of this formulation, and the developed analytical method was applied to monitor the content of active ingredients. Interestingly, these investigations led to the detection of a known clotrimazole degradation product whose structure was confirmed using NMR and HRMS, as well as a possible interaction between itraconazole and benzyl alcohol.

Keywords: pre-development process; clotrimazole; itraconazole; stability; method validation; sporotrichosis

1. Introduction

Sporotrichosis is a subcutaneous infectious disease with subacute to chronic evolution and with a worldwide distribution. The etiologic agent of sporotrichosis is *Sporothrix schenckii*, which is a thermo-dimorphic fungus that lives saprophytically in nature and is pathogenic to humans and animals [1,2]. The occurrence of sporotrichosis in animals, especially cats, as well as its transmission to humans has been reported in several countries [3]. In this context, the Brazilian state of Rio de Janeiro is an epidemic area for this disease and the first one associated with zoonotic transmission related to sick felines by *Sporothrix brasiliensis*, the most virulent species from the *S. schenckii* complex [4].

The treatment of both feline and human sporotrichosis is based on the use of itraconazole 1, which contains the 1,2,4-triazole scaffold in its structure and inhibits the synthesis of sterol, a vital component of the fungus cell membrane [5,6]. Clotrimazole 2, on the other hand, is an imidazole derivative with antifungal activity that is only indicated for topical use due to its toxicity (Figure 1). Similarly to itraconazole, clotrimazole is a synthetic antifungal and its mechanism of action involves the inhibition of sterol biosynthesis [7]. In this sense, Gagini et al. [8] reported the effectiveness of the combination of itraconazole with clotrimazole against *S. brasiliensis* yeasts (the infective form) from feline and human sporotrichosis isolates, suggesting that clotrimazole by itself or in combination with itraconazole is potentially a new option for the treatment of sporotrichosis.

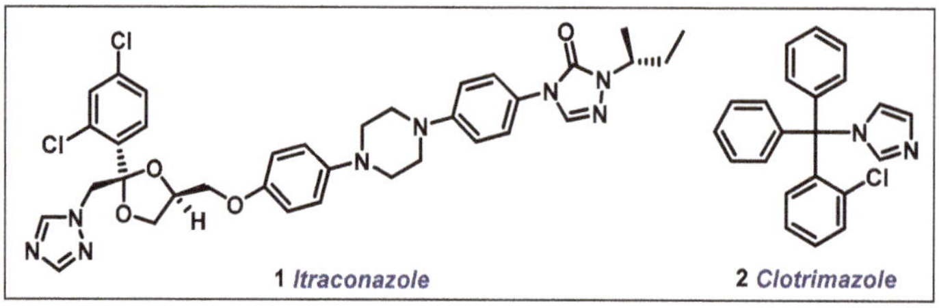

Figure 1. Chemical structures of clotrimazole and itraconazole.

Accordingly, the development of new pharmaceutical technologies for the use of clotrimazole and itraconazole associations is highly desirable in order to increase their efficiency in therapy, decrease adverse effects and provide, especially for felines, alternative treatments. Moreover, the use of a combination antifungal therapy is a promising approach to avoid resistance [9]. Allied to all the mentioned features, the association of known drugs is highly advantageous for the pharmaceutical industry to find innovations for the market, since they can reformulate their products in a more economically advantageous way when compared to the development of new drugs. In addition, the association of drugs already in use in the pharmaceutical market may increase their efficiency with known safety and effectiveness, reintroducing forgotten and/or discarded ones.

Considering the development of new formulations, microemulsions (MEs) have attracted great interest as potential drug delivery systems, mainly due to their unique physicochemical properties such as drug solubilization and enhanced absorption properties [10,11]. MEs are a thermodynamically stable, isotropic, transparent liquid system consisting of two immiscible liquids (usually water and oil) stabilized by a film of surfactant compounds, suitably combined with a cosurfactant [12,13]. The presence of the surfactant helps to reduce the interfacial tension, making it possible to join the oil and aqueous phases [14,15]. MEs have been proposed as an innovative formulation approach to improve solubility and efficacy and reduce of the toxicity of various drugs. Therefore, when the known hydrophobicity of clotrimazole and itraconazole are taken into account, such systems could be particularly advantageous for their delivery.

In light of the aforementioned concepts, this paper reports the initial research phase for the pre-development of a clotrimazole–itraconazole formulation, the first step towards a new antifungal combination. In this sense, the development and characterization of this new pharmaceutical formulation requires the evaluation of parameters such as drug release and stability. Therefore, as a further extension of our work in the field, we have developed a simple, sensitive, and specific HPLC method for the simultaneous quantification of clotrimazole and itraconazole in microemulsion. Although many researchers have investigated clotrimazole and itraconazole singly or in combination with other compounds, to the best of our knowledge, no HPLC method has been developed for the simultaneous determination of both drugs simultaneously, especially in microemulsion systems [16,17]. Finally, we performed an accelerated stability study of this formulation and the developed analytical method was applied to monitor the content of active ingredients. Interestingly, these investigations led to the detection of a known clotrimazole degradation product whose structure was confirmed using NMR and HRMS, as well as a possible interaction between itraconazole and benzyl alcohol.

2. Results and Discussion

2.1. Study of the Compatibility between Clotrimazole and Itraconazole

We initiated our studies by analyzing the physicochemical properties of both active ingredients as well as their compatibility using different techniques such as thermal analyses (differential scanning calorimetry (DSC) and thermogravimetric/derivative thermogravimetry (TG)/DTG) analysis), powder X-ray diffraction (XRD) and FTIR.

Initially, we proceeded to characterize the active ingredients and their combination using thermal analyses, which offer the ability to quickly screen for potential drug–drug incompatibilities. Such interactions can be of a physical or chemical nature and may affect the stability and bioavailability of the final product, compromising the therapeutic efficacy and safety [18].

The TG and DTG curves of clotrimazole (Figure 2a) showed that it is thermally stable up to 340 °C, when its thermal decomposition starts; the highest rate of weight loss occurs at 388.6 °C, as showed in the DTG curve, and is finished at 421.1 °C, where a loss of 60% of the total weight is observed. As for itraconazole, its thermal decomposition starts at 200 °C and is finished at 348.4 °C, with a maximum rate at 295.3 °C and a total weight loss of 87%. The TG profile of the binary mixture of clotrimazole and itraconazole (1:1 ratio) showed two decomposition steps, indicating that the compounds undergo thermal degradation independently, although a small shift in the initial temperature of decomposition was observed, as expected.

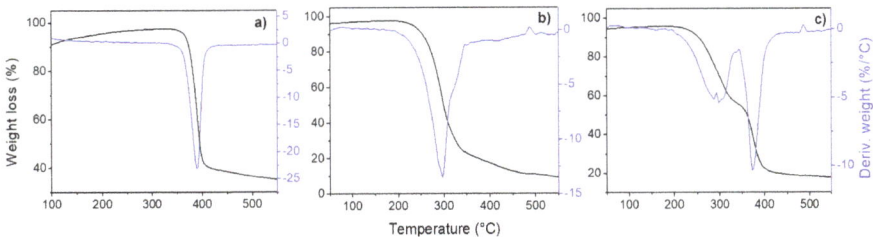

Figure 2. Thermogravimetric (TG) and derivative thermogravimetry (DTG) curves for (**a**) clotrimazole, (**b**) itraconazole and (**c**) the binary mixture of clotrimazole and itraconazole (1:1).

Next, the DSC technique was employed to further analyze the occurrence of events related to possible interactions between the drugs [19]. It is noteworthy that although such analyses are conducted upon heating the sample to high temperatures, which is not consistent with the process of drug production nor its administration to patients, they afford important information regarding the physical properties of the sample [18].

The DSC curves of the drugs showed endothermic peaks attributed to the melting of the drugs between 158.5 and 175.0 °C ($\Delta H = 31.5$ J g^{-1}) for itraconazole and 136.8 and 153.1 °C ($\Delta H = 41.6$ J g^{-1}) for clotrimazole. On the other hand, a single endothermic event was observed in the DSC curve of the binary mixture, starting at 127.7 and finishing at 137.1 ($\Delta H = -25.35$ J g^{-1}), which suggests a strong interaction between clotrimazole and itraconazole (Figure 3).

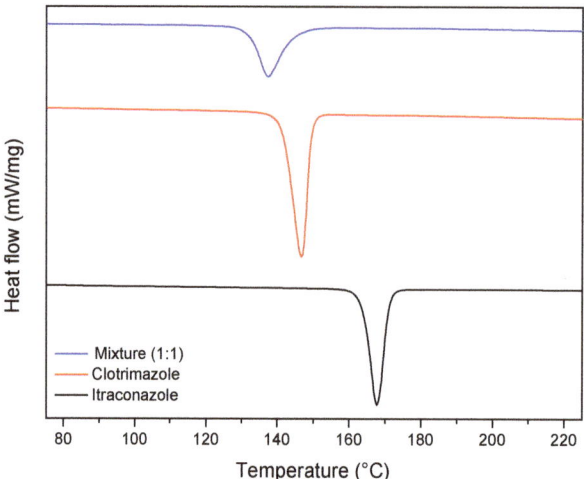

Figure 3. Differential scanning calorimetry (DSC) profile of itraconazole, clotrimazole, and the clotrimazole/itraconazole binary mixture (1:1).

In order to further explore the possibility of interactions between the active ingredients, powder X-ray diffraction (PXRD) analyses were conducted. Interestingly, the diffractogram of the binary mixture (Figure 4) contained virtually all the peaks of clotrimazole and itraconazole, with no marked displacement of the peaks being observed. Furthermore, it is important to highlight that it was not possible to notice the appearance of any new peaks, which means that if there is any interaction between the drugs, it probably is not strong enough to take place in the solid state. The same observations were made in the FTIR spectra of the binary mixture, which showed the characteristic bands observed for the isolated active ingredients (For more details, see the Supplementary Materials).

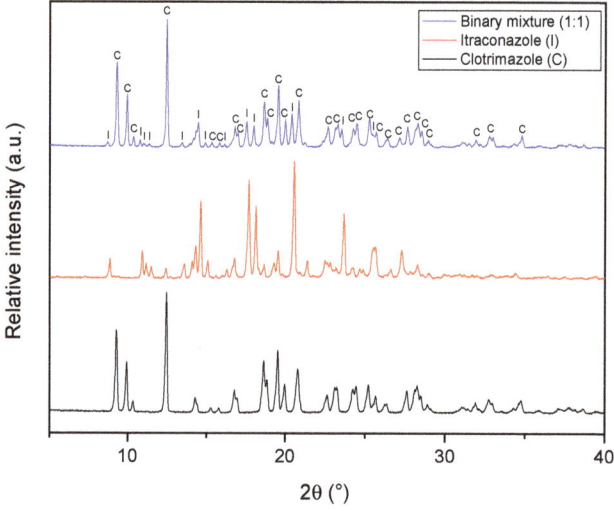

Figure 4. X-ray diffractograms of clotrimazole, itraconazole, and the binary mixture (1:1).

With the characterization of the active ingredients and the binary mixture in hand, we proceeded to develop an HPLC method for their quantification in a newly developed microemulsion for the treatment of sporotrichosis.

2.2. Determination of the Concentration of Clotrimazole and Itraconazole in Microemulsions Using HPLC Analyses

Considering the unique properties presented by microemulsions, in the present work, benzyl alcohol was used as an oil phase, Tween® 60 as a surfactant, and propylene glycol as a cosolvent in the presence of water. These components were chosen on the basis in their previously reported applications in other pharmaceutical forms available on the international market.

In this context, HPLC-DAD (diode array detector) was selected as an analytical tool for the simultaneous quantification of clotrimazole and itraconazole in the developed microemulsion through a rapid, simple, and isocratic method [20]. In our study, the best separation condition was achieved using a C18 analytical column with a mobile phase composed of acetonitrile and a phosphate buffered saline 0.05 M (pH 8.0 with ammonium hydroxide 1 M) in the ratio (v/v) 60:40, respectively, with a 1 mL min^{-1} flow rate and UV detection at 190 nm. A typical chromatogram is presented in Figure 5, with a retention time of 9.1 min being observed for clotrimazole and 10.9 min for itraconazole.

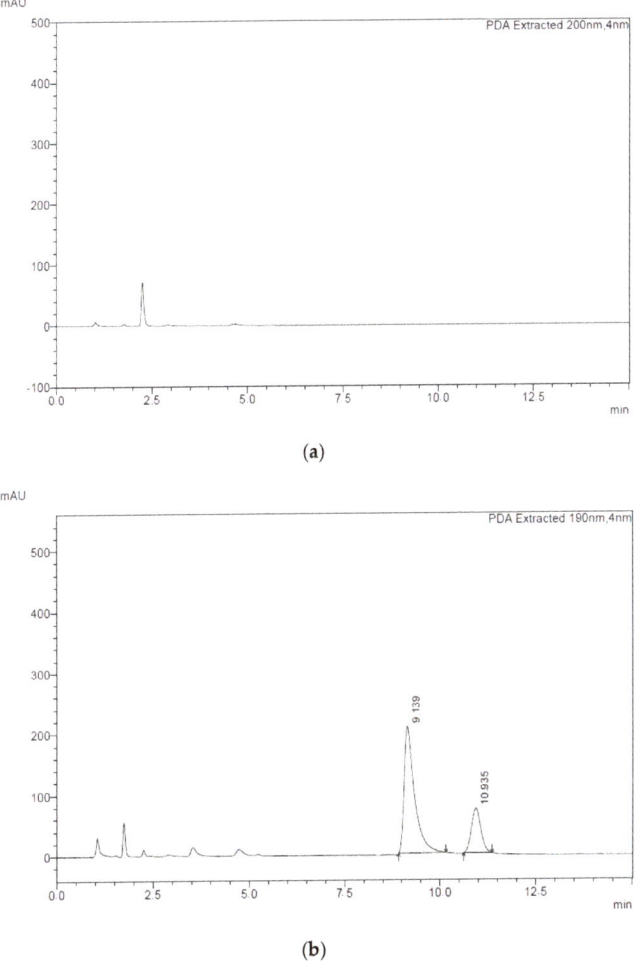

Figure 5. Chromatograms of the (a) mobile phase and (b) standard solution containing a binary mixture of itraconazole and clotrimazole.

To evaluate the linearity of the method, calibration standards of clotrimazole (5–200 µg mL^{-1}) and itraconazole (5–160 µg mL^{-1}) were analyzed. A linear relationship was established for the injected concentration ranges versus the peak area for both analytes, with determination coefficients greater than 0.9988 (see the calibration curves in the Supplementary Materials). The calibration curve parameters are reported in Table 1, with the linearity parameters of the method shown in Table 2.

Table 1. Summary of the validation data obtained for the proposed HPLC method developed for the quantification of clotrimazole and itraconazole in microemulsions. LOD—limit of detection; LOQ—limit of quantification.

Standard Solutions	Parameters of the Method	Validation Results
Clotrimazole	Linearity	Calibration range (µg mL^{-1}): 5–200 y = 233647.7939x − 312039.9299 (R^2 = 0.9988)
	LOD	0.84 µg mL^{-1}
	LOQ	2.54 µg mL^{-1}
	Slope	233647.7939 ± 976.8015153
	Interception	−312039.9299 ± 59416.57811
Itraconazole	Linearity	Calibration range (µg mL^{-1}): 5–160 y = 89946.6896x − 79996.5373 (R^2 = 0.9999)
	LOD	0.86 µg mL^{-1}
	LOQ	2.60 µg mL^{-1}
	Slope	89946.6896 ± 780.1420761
	Interception	−79996.53731 ± 23351.48986

Table 2. Data related to the linearity of the developed HPLC method with its respective average, precision, and accuracy.

Concentration (µg/mL)	Clotrimazole			Itraconazole		
	Average (µg/mL)	Accuracy (%)	Precision (%)	Average (µg/mL)	Accuracy (%)	Precision (%)
5	4.883	97.7	0.20	5.593	111.9	0.86
10	9.292	92.9	0.57	10.115	101.2	0.91
20	19.233	96.2	0.40	20.029	100.1	0.58
40	38.927	97.3	0.01	39.621	99.1	1.12
80	77.731	97.2	1.08	79.154	98.9	1.81
160	151.888	94.9	0.71	160.488	100.3	0.82
200	204.631	102.3	0.65	-	-	-

The method's selectivity was confirmed by the absence of interferences at the retention times of itraconazole and clotrimazole in the microemulsion prepared without the drugs (Figure 6). The purity of the compounds was checked using PDA (photodiode array) detection. The within-assay precision (repeatability) was carried out by performing six consecutive analyses of standard solution at three different concentrations for each drug on the same day. The samples were also analyzed on different days to evaluate the between-assay precision (intermediate precision). The obtained values were evaluated through the dispersion of the results by calculating the standard deviation of the measurement series. The intra- and inter-day precision relative standard deviation (RSD %) was between 1.18 and 0.8 for clotrimazole and 1.48 and 0.84 for itraconazole. The recovery of the drugs was in the range of 93.8–100.9% with RSDs below 2.35% for clotrimazole and in the range of 100.5–104.3% with RSDs below 2.40% for itraconazole. The results are given in Table 3.

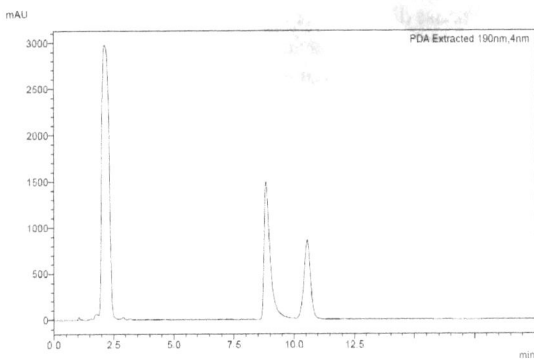

Figure 6. Chromatogram obtained from the injection of the microemulsion using the developed HPLC method.

Table 3. Data related to the repeatability and intermediate precision of the developed HPLC method.

Samples (µg mL^{-1})	Intra-Day Precision (Repeatability)			Inter-Day Precision (Intermediate Precision)		
Clotrimazole	Concentration Found (µg mL^{-1})	Accuracy (%)	Precision (%)	Concentration Found (µg mL^{-1})	Accuracy (%)	Precision (%)
7	6.818	97.4 ±2.25	0.47	6.865	98.07 ± 1.17	2.35
15	14.510	96.7 ±1.13	1.18	14.075	93.83 ± 3.17	0.95
120	116.679	97.2 ±0.27	0.28	121.108	100.92 ± 4.28	0.28
Itraconazole	Concentration Found (µg mL^{-1})	Accuracy (%)	Precision (%)	Concentration Found (µg mL^{-1})	Accuracy (%)	Precision (%)
7	7.206	102.9 ± 1.33	1.48	7.305	104.35 ± 1.25	1.20
70	70.809	101.2 ± 1.15	1.16	70.374	100.53 ± 2.41	2.40
150	152.745	101.8 ± 0.85	0.84	160.98	100.61 ± 4.9	1.59

No changes were observed in the drug concentrations of the stock solutions under storage conditions. Indeed, further analyses showed that the percent recovery of clotrimazole and itraconazole were, respectively, 97.3% ± 3.15 and 91.3% ± 2.71 at room temperature (25 °C) and 94.2 ± 0.34 and 88.7 ± 1.63 under refrigeration (−5 °C, Table 4). Moreover, the drugs were stable for at least 30 days under storage conditions, with RSDs below 8%.

Table 4. Data related to the stability of the assay of the developed HPLC method. N = 2 for each day and condition.

Days	Accuracy (%)	Precision (%)	Accuracy (%)	Precision (%)
	Clotrimazole		Itraconazole	
0	97.3 ± 0.94 (25 °C)	1.18	101.4 ± 0.62	0.84
7	105.7 ± 0.89 (25 °C)	0.85	98.6 ± 4.48	4.57
	104.9 ± 0.07 (−5 °C)	0.07	98.4 ± 7.79	7.96
15	105.3 ± 1.51 (25 °C)	1.45	101.3 ± 0.51	0.50
	105.5 ± 0.39 (−5 °C)	0.38	100.1 ± 3.71	3.74
30	97.3 ± 3.15 (25 °C)	0.62	91.3 ± 2.71	3.21
	94.2 ± 0.34 (−5 °C)	0.73	88.7 ± 1.63	3.04

In order to evaluate the robustness of the chromatographic method, assays were carried out by changing both the column brand and ratio of the mobile phase for acetonitrile 70:30 (v/v) and a phosphate buffered saline 0.05 M (pH 8.0 with ammonium hydroxide 1 M). The alteration of the column brand and the mobile phase did not promote any significant variations in the retention time of clotrimazole and itraconazole peaks; a good resolution was observed with retention times of 8 min for clotrimazole and 10.7 min for itraconazole (Figure 7).

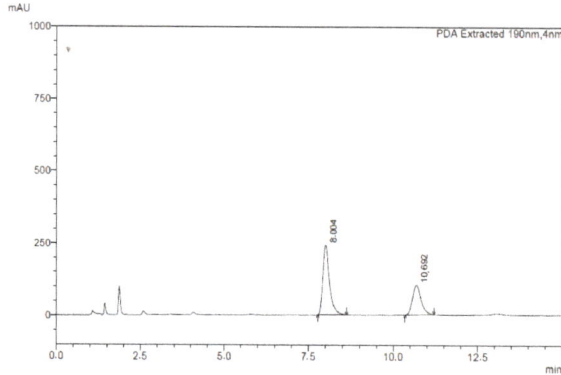

Figure 7. Chromatogram of clotrimazole and itraconazole obtained in the robustness studies.

2.3. *Study of the Stability of a Novel Microemulsion Containing Clotrimazole and Itraconazole*

Subsequently, the developed method was used in the determination of clotrimazole and itraconazole in the newly developed microemulsion with the purpose of quantifying the drugs in the formulation, as well as in the accelerated stability study. Based on the assumption that possible interactions and incompatibilities may arise from the contact between the drugs over time, they were left to stand for three months, both under refrigeration and heating conditions, and further analyzed.

The initial drug content of the microemulsion was taken as 100%, and the drug content over time was plotted (Figure 8), with all data being represented as mean ± SD (n = 3). For the samples stored at 5 °C, no significant changes were observed for both drugs when compared to the first day. Furthermore, it is noteworthy that there was no evident interaction between clotrimazole and itraconazole at this temperature, since the peaks of both drugs were detected independently without the appearance of any additional peaks. On the other hand, when the samples that were stored at 40 °C were analyzed, it was possible to notice a significant decrease in the concentration of the drugs over time, especially for clotrimazole. Additionally, a new peak could also be observed in the chromatogram of such samples (Figure 9).

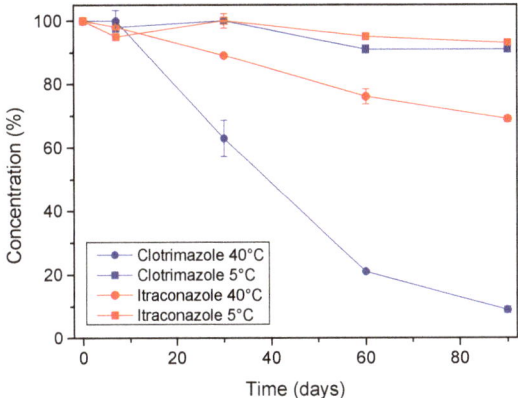

Figure 8. Graph showing the concentration of clotrimazole and itraconazole over time in different conditions. All data is represented as mean ± SD (n = 3).

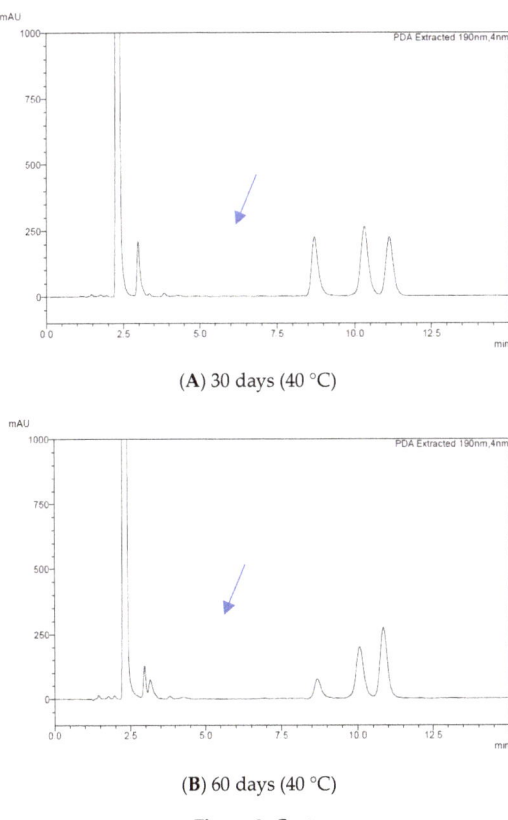

Figure 9. Cont.

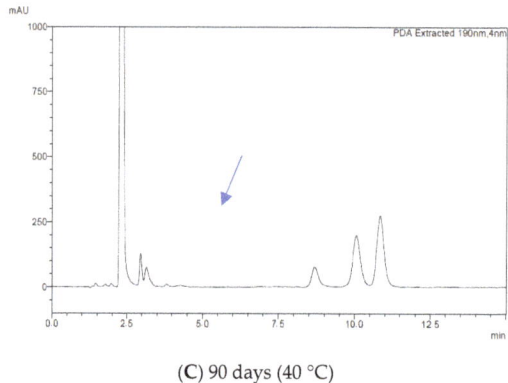

(**C**) 90 days (40 °C)

Figure 9. HPLC chromatograms for the samples in the stability study after (**A**) 30 days, (**B**) 60 days, and (**C**) 90 days.

In order to investigate the formation of this compound, which might be a result of the interaction between clotrimazole and itraconazole, we conducted further studies. Initially, we sought to investigate which degradation products could be formed from the degradation of both drugs and found that the degradation of clotrimazole is well-reported under acidic conditions, giving product **3** (Figure 10).

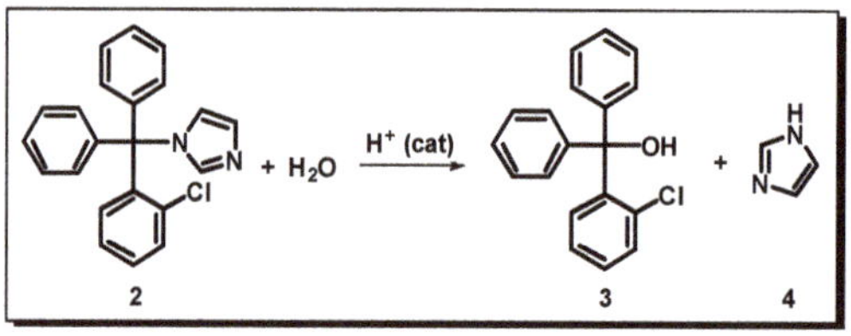

Figure 10. Reaction scheme showing the degradation of clotrimazole in acid medium.

With these concepts in mind, we conducted the synthesis of compound **3** from clotrimazole by heating it at 80 °C in the presence of acetonitrile and concentrated hydrochloric acid for 2 h; the product identity was confirmed using NMR and HRMS by comparing the obtained data with previous reports (for details, see the Supplementary Materials) [21]. Next, we conducted the forced degradation of a mixture of itraconazole and clotrimazole by heating both at 50 °C for 24 h in a solution of acetonitrile, water, and benzyl alcohol-mimicking the microemulsion composition—and isolated the formed product using column chromatography.

With both compounds in hand, we analyzed product **3** and the degradation product by HPLC using the developed method, and the comparison of the retention times of both compounds proved that, indeed, product **3** is formed from the degradation of clotrimazole under acidic conditions. Furthermore, the retention time was also a match for the product previously detected in the stability studies conducted at 40 °C, which proves that under specific conditions, clotrimazole may undergo degradation in the presence of traces of acid, forming **3**. However, the formation of **3** was not observed in the stability studies conducted at 5 °C, which shows the viability of this novel microemulsion and encourages carrying out further studies for its development.

The content of itraconazole (% w/w) in the microemulsions stored in climatic chambers also underwent a slight decrease, which was less significant when compared to clotrimazole. In order to exclude the possibility of interaction between the drugs, the decrease in itraconazole content was also investigated. However, unlike clotrimazole, degradation studies of itraconazole are not found in the literature. Thus, an aliquot was collected directly from the chromatographic system at the same retention time as the degradation product formed during the stability study; at a low intensity, it was possible to observe a product with a mass-to-charge ratio (*m/z*) of 437.1931. Considering that the itraconazole concentration change was lower than for clotrimazole, we hypothesized that an interaction of itraconazole with some other excipient of the microemulsion may be taking place. In that sense, we propose that the degradation product may be formed by the nucleophilic addition of benzyl alcohol to the methylene group linking the phenolic aromatic part with the 1,3-dioxolane ring (Figure 11); indeed, the mass-to-charge ratio was a match for the proposed product. It is worth mentioning that although we have observed a good match in a mass-to-charge ratio of 437.1931, further studies are necessary to confirm whether the proposed structure is indeed the correct one, such as the isolation and complete spectroscopic characterization of this compound, which was not possible at the scale we were working.

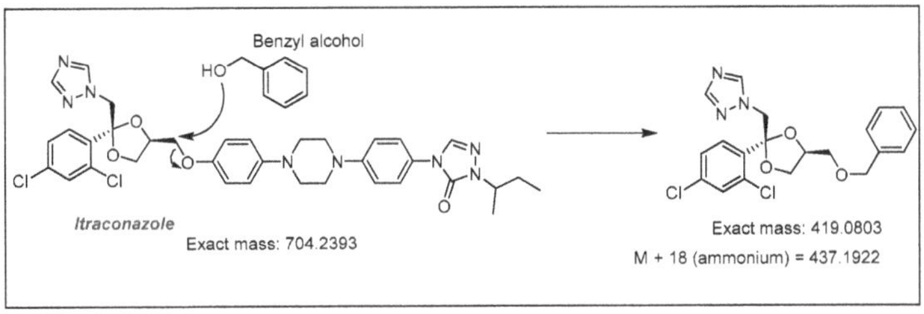

Figure 11. Scheme showing the reaction between itraconazole and benzyl alcohol.

3. Experimental Methods

3.1. Materials for Analytical Method Development

Clotrimazole and itraconazole (as a mixture of stereoisomers) standards were purchased from Merck, São Paulo, SP, Brazil. Microemulsions were prepared using Tween® 60, propylene glycol, and benzyl alcohol, all purchased from Merck, São Paulo, SP, Brazil. HPLC-grade acetonitrile was acquired from J.T. Baker Inc., Phillipsburg, NJ, USA. Clotrimazole (Jintan Zhongxing Pharmaceutical Chemical Co., Ltd., Mainland, China) and itraconazole (Metrochem API, Telangana, India) were donated by Valdequimica Produtos Quimicos Ltd., São Paulo, Brazil. All solutions were prepared with ultra-pure Milli-Q water obtained from a Milli-Q Water Millipore purification system (Burlington, MA, USA).

3.2. Compatibility Study of Clotrimazole and Itraconazole

3.2.1. Preparation of Clotrimazole/Itraconazole Binary Mixtures

The binary mixtures were prepared and homogenized by taking clotrimazole and itraconazole in a 1:1 proportion (*w:w*). These mixtures were further used for X-ray powder diffraction, Fourier transform infrared spectroscopy (FTIR), and thermal analyses.

3.2.2. X-ray Powder Diffraction (PXRD)

PXRD patterns were collected on a Bruker D8 Venture diffractometer system (Bruker, Billerica, MA, USA) operating at 1.5406 Å, 40 kV voltage, and a current of 40 mA using a Cu Kα radiation source. The samples were contained in a flat poly(methyl methacrylate) sample holder and the data acquisition was done in a range of 5 to 70° (2θ) at 0.019°/0.1 s step size over a total period of 10 min.

3.2.3. Fourier Transform Infrared Spectroscopy (FTIR)

The FTIR spectra of the solid samples were obtained using a Varian FT-IR 660 equipment (Varian Inc., Walnut Creek, CA, USA). A hydraulic press was used to prepare pellets for analysis. The KBr pellets contained 3 mg of a sample and 100 mg of KBr. Spectra were collected with a resolution of 4 cm^{-1} on the spectral domain of 3800–600 cm^{-1}.

3.2.4. Thermal Analyses

DSC data were collected on a Shimadzu Differential Scanning Calorimeter DSC-60A (Shimadzu, Quioto, Japan). Approximately 4 mg samples were placed in aluminum pans, and the temperature program was set to increase from 30 to 250 °C with a heating rate of 10 °C min^{-1} under nitrogen flow (50 mL min^{-1}).

Thermogravimetric (TG) analyses were performed using a Netzsch STA 409 PC/PG (Netzsch, Selb, Germany) under a nitrogen atmosphere with a flow rate of 60 mL min^{-1} at a heating rate of 10 °C min^{-1} over the range of 30 to 300 °C and using 6 mg of sample in an aluminum cell.

3.3. Instruments and Chromatographic Conditions

Chromatographic experiments were performed on a Shimadzu SPD-M20A system (Shimadzu, Quioto, Japan). The chromatographic separations were performed using a 150 mm × 4.6 mm i.d. (5 μm particle size) Fortis C18 column in isocratic elution mode with acetonitrile and phosphate buffered saline 0.05 M pH 8.0 adjusted with ammonium hydroxide 1 M (60:40, *v/v*) at a flow rate of 1.0 mL min^{-1}. The detection wavelength was set at 190 nm, and the injection volume was 20 μL.

3.4. Standard Stock Solutions and Calibration Standards

Standard stock solutions of clotrimazole and itraconazole were freshly prepared by dissolving the drugs in methanol (0.2 mg mL^{-1}) Calibration standards in the concentration range of 5, 10, 20, 40, 80, 160, and 200 μg mL^{-1} were prepared in the appropriate volumetric flasks by diluting the stock solution in the mobile phase. An aliquot (20 μL) of the solution was then directly injected into the HPLC.

3.5. Sample Preparation

An amount of microemulsion was accurately weighted to contain 25 mg clotrimazole and itraconazole in a 50 mL centrifuge tube and heated for 5 min in a water bath at 50 °C. The sample was then removed from the bath, shaken until cooled to room temperature, and placed in an ice-methanol bath. Next, the sample was centrifuged for 5 min and extracted with chloroform (5 mL). Finally, the solvent was removed under a stream of gaseous nitrogen, and the residue was diluted in the mobile phase.

3.6. Method Validation Protocol

The proposed method was validated under the optimized conditions regarding its linearity range, selectivity, sensitivity, precision, accuracy and stability of the assay according to the regulatory guidelines requirements (FDA).

3.6.1. Linearity Range

The linearity range was evaluated by measuring the chromatographic peak area responses of the drugs at seven concentration levels and in triplicate. Analytical curves were constructed by plotting the peak area against the concentration of itraconazole and clotrimazole (Figures 2 and 3), which gives the regression equation. The results are presented in Table 1.

3.6.2. Selectivity

To ensure the selectivity of the proposed method, drug-free microemulsions were prepared and analyzed in the described chromatographic conditions.

3.6.3. Sensitivity

The sensitivity was determined by means of the limit of detection (LOD) and limit of quantification (LOQ). One of the ways to calculate the LOD (Equation (1)) and LOQ (Equation (2)) is based on the standard deviation (σ) of the y-intercept from the regression of the calibration standard. The results are given in Table 1.

$$LOD = \frac{3,3.\sigma}{s} \quad (1)$$

LOD (σ—standard deviation; s—slope of the calibration standard).

$$LOQ = \frac{10.\sigma}{s} \quad (2)$$

LOQ (σ—standard deviation; s—slope of the calibration standard).

3.6.4. Precision and Accuracy

The accuracy and precision of the method were estimated by quintuplicate quality control (QC) samples prepared using the mobile phase: 7 µg mL^{-1} (low QC), 15 µg mL^{-1} (medium QC), and 120 µg mL^{-1} (high QC) for clotrimazole and 7 µg mL^{-1} (low QC), 70 µg mL^{-1} (medium QC), and 150 µg mL^{-1} (high QC) for itraconazole. Accuracy was established through back-calculation and expressed as the percent difference between the found and the nominal concentration for each compound, and the precision was calculated as the coefficient of variation (CV) of the replicate measurements. Calibration standards and QC samples were analyzed in three different batches in order to determine the intra and inter-batch variability.

3.6.5. Stability

The stability of the standard solutions was investigated after storage for 7, 15, and 30 days at room temperature (25 °C) and under refrigeration (−5 °C) using the working solution.

3.6.6. Robustness

The robustness of an analytical method is a measure of its capacity to resist changes due to small variations in parameter conditions, e.g., by using a different column. In this way, the method robustness was assessed as a function of changing the column brand for a C18 Agilent column (Agilent Technologies Inc, Santa Clara, CA, USA), (150 × 4.6 mm × 5 µm) and the ratio of the mobile phase.

3.7. Application of the Method

3.7.1. Microemulsion Preparation

With the developed method in hand, the next step was to develop a stable microemulsion using a combination of clotrimazole and itraconazole. MEs were composed of benzyl alcohol, the non-ionic surfactant Tween® 60, propylene glycol, and water. The optimum weight ratios of the components and MEs' areas were determined using a pseudo-ternary phase diagram (data not shown in this work). The systems were prepared as previously described [22]; the surfactant (Tween® 60) and cosolvent (propylene glycol) were prepared separately, and clotrimazole and itraconazol were solubilized in benzyl alcohol and added to the mixture. The pseudo-ternary phase diagrams of oil, surfactant/cosolvent, and water were set up using the water titration method.

3.7.2. Stability Study

The stability profile of the prepared microemulsion at accelerated conditions was studied according to the ICH guidelines. The formulation was placed separately in an amber-colored screw-capped glass container and stored at 40 ± 2 °C and 5–8 ± 3 °C for 3 months, with sampling at 0, 30, 60, and 90 days. The samples were then evaluated for drug content using the developed HPLC method.

3.8. Characterization of the Synthetized Compounds

NMR spectra were obtained using a Varian Unity Plus VXR (Varian Inc., Walnut Creek, CA, USA), 500 MHz instrument in CDCl$_3$ solutions. The chemical shifts were reported in units of d (ppm) downfield from tetramethylsilane, which was used as an internal standard; coupling constants (J) are reported in hertz and refer to apparent peak multiplicities. High-resolution mass spectra (HRMS) were recorded on a MICROMASS Q-TOF mass spectrometer (Waters, Milford, MA, USA).

4. Conclusions

The combination of clotrimazole and itraconazole in a pharmaceutical formulation is of great importance owing to the potential of generating a new option for the treatment of sporotrichosis. In this sense, the preformulation investigation using different techniques (DSC, TG, PXRD, FTIR) was essential to examine the existence of possible clotrimazole–itraconazole interactions.

Furthermore, an HPLC method was developed and validated according to standard guidelines, and it is the first reported method for the simultaneous determination of clotrimazole and itraconazole in nanotechnology-based products such as microemulsions. Based on our results, it was possible to conclude that there is no other co-eluting peak along with those of interest, the method being specific for the estimation of clotrimazole and itraconazole.

Interestingly, accelerated stability studies showed that a product derived from clotrimazole was formed, as well as a possible interaction between itraconazole and benzyl alcohol, when the microemulsion was conditioned at elevated temperatures (40 °C). On the other hand, the studies conducted at 5 °C showed that the microemulsion is stable for at least 3 months, as no degradation peaks were observed in the HPLC analysis, which allows us to infer that it is possible to guarantee the stability of the formulation under refrigeration.

Supplementary Materials: Supplementary materials are available online. Figure S1: Analytical calibration curve for clotrimazole, Figure S2: Analytical calibration curve for itraconazole, Figure S3: IR spectra of clotrimazole, itraconazole and their binary mixture (1:1), Figure S4: HPLC chromatogram of compound 3, Figure S5: HPLC chromatogram of the decomposition product formed via the forced degradation of clotrimazole in the presence of itraconazole.

Author Contributions: All authors have read this manuscript and concur with its submission. The contributions of each author are listed as follows: P.G.F.—Conceptualization, Data curation, Formal analysis, Investigation, Methodology, Roles/Writing—original draft; C.G.d.S.L.—Conceptualization, Data curation, Formal analysis, Investigation, Methodology, Writing—review & editing; L.L.N.—Data curation, Formal analysis, Investigation; Methodology; M.C.d.M.—Conceptualization, Data curation, Formal analysis, Investigation, Methodology, Funding acquisition, Writing—review & editing; F.d.C.d.S.—Conceptualization, Funding acquisition, Project administration, Resources, Supervision, Validation, Visualization; A.L.V.—Conceptualization; D.O.F.—Conceptualization, Funding acquisition, Supervision, Writing—review & editing; V.F.F.—Conceptualization, Funding acquisition, Project administration, Resources, Supervision, Writing—review & editing.

Funding: This study was financed in part by the Coordenação de Aperfeiçoamento de Pessoal de Nível Superior-Brasil (CAPES)-Finance Code 001, CNPq (303713/2014-3) and FAPERJ (E-26/2002.800/2017, E-26/200.930/2017).

Conflicts of Interest: All authors declare that there is no conflict of interest.

References

1. Chakrabarti, A.; Bonifaz, A.; Gutierrez-Galhardo, M.C.; Mochizuki, T.; Li, S. Global epidemiology of sporotrichosis. *Med. Mycol.* **2015**, *53*, 3–14. [CrossRef] [PubMed]
2. Gremião, I.D.; Miranda, L.H.; Reis, E.G.; Rodrigues, A.M.; Pereira, S.A. Zoonotic epidemic of sporotrichosis: Cat to human transmission. *PLoS Pathog.* **2017**, *13*, e1006077. [CrossRef] [PubMed]
3. Rodrigues, A.M.; de Hoog, G.S.; de Camargo, Z.P. *Sporothrix* species causing outbreaks in animals and humans driven by animal-animal transmission. *PLoS Pathog.* **2016**, *12*, e1005638. [CrossRef] [PubMed]
4. Barros, M.B.L.; Schubach, T.P.; Coll, J.O.; Gremião, I.D.; Wanke, B.; Schubach, A. Esporotricose: A evolução e os desafios de uma epidemia. *Rev. Panam. Salud Publica* **2010**, *27*, 455–460. [PubMed]

5. Gremião, I.D.; Menezes, R.C.; Schubach, T.M.; Figueiredo, A.B.; Cavalcanti, M.C.; Pereira, S.A. Feline sporotrichosis: Epidemiological and clinical aspects. *Med. Mycol.* **2015**, *53*, 15–21. [CrossRef] [PubMed]
6. Bustamante, B.; Campos, P.E. Sporotrichosis: A forgotten disease in the drug research. *Expert Rev. Anti-Infect. Ther.* **2004**, *2*, 85–94. [CrossRef] [PubMed]
7. Kadavakollu, S.; Stailey, C.; Kunapareddy, C.S.; White, S. Clotrimazole as a cancer drug: A short review. *Med. Chem.* **2014**, *4*, 722–724. [CrossRef]
8. Gagini, T.; Borba-Santos, L.P.; Rodrigues, A.M.; Camargo, Z.P.; Rozental, S. Clotrimazole is highly effective in vitro against feline *Sporothrix brasiliensis* isolates. *J. Med. Microbiol.* **2011**, *66*, 1573–1580. [CrossRef]
9. Pai, V.; Ganavalli, A.; Kikkeri, N.N. Antifungal resistance in dermatology. *Indian J. Dermatol.* **2018**, *63*, 361–368. [CrossRef]
10. Carvalho, A.L.M.; da Silva, J.A.; Lira, A.A.M.; Conceição, T.M.F.; Nunes, R.S.; Junior, R.L.C.A.; Sarmento, V.H.V.; Leal, L.B.; Santana, D.P. Evaluation of microemulsion and lamellar liquid crystalline systems for transdermal zidovudine delivery. *J. Pharm. Sci.* **2016**, *105*, 1–6. [CrossRef]
11. Padula, C.; Telò, I.; Ianni, A.D.; Pescina, S.; Nicoli, S.; Santi, P. Microemulsion containing triamcinolone acetonide for buccal administration. *Eur. J. Pharm. Sci.* **2018**, *115*, 233–239. [CrossRef]
12. Rashida, M.A.; Naza, T.; Abbasa, M.; Nazirb, S.; Younasa, N.; Majeeda, S.; Qureshic, N.; Akhtard, M.N. Chloramphenicol loaded microemulsions: Development, characterization and stability. *Colloid Interface Sci. Commun.* **2019**, *28*, 41–48. [CrossRef]
13. Seok, S.H.; Lee, S.-A.; Park, E.-S. Formulation of a microemulsion-based hydrogel containing celecoxib. *J. Drug Deliv. Sci. Technol.* **2018**, *43*, 409–414. [CrossRef]
14. Kumar, S.K.; Dhancinamoorthi, D.; Sarvanan, R.; Gopal, U.K.; Shanmugam, V. Microemulsions as a carrier for novel drug delivery: A review. *Int. J. Pharm. Sci. Rev. Res.* **2011**, *10*, 37–45.
15. Hu, X.-B.; Kang, R.-R.; Tang, T.-T.; Li, Y.-J.; Wu, J.-Y.; Wang, J.-M.; Liu, X.-Y.; Xiang, D.-X. Topical delivery of 3,5,4′-trimethoxy-trans-stilbene-loaded microemulsion-based hydrogel for the treatment of osteoarthritis in a rabbit model. *Drug Deliv. Transl. Res.* **2019**, *9*, 357–365. [CrossRef]
16. Hájková, R.; Sklenárová, H.; Matysová, L.; Svecová, P.; Solich, P. Development and validation of HPLC method for determination of clotrimazole and its two degradation products in spray formulation. *Talanta* **2007**, *73*, 483–489. [CrossRef]
17. Abdel-Moety, E.M.; Khattab, F.I.; Kelani, K.M.; AbouAl-Alamein, A.M. Chromatographic determination of clotrimazole, ketoconazole and fluconazole in pharmaceutical formulations. *Farmaco* **2002**, *57*, 931–938. [CrossRef]
18. Bharate, S.S.; Bharate, S.B.; Bajaj, A.N. Interactions and incompatibilities of pharmaceutical excipientes with active pharmaceutical ingredients: A comprehensive review. *J. Excipients and Food Chem.* **2010**, *1*, 3–26. [CrossRef]
19. Ceschel, G.C.; Badiello, R.; Ronchi, C.; Maffei, P. Degradation of components in drug formulations: A comparison between HPLC and DSC methods. *J. Pharm. Biomed. Anal.* **2003**, *32*, 1067–1072. [CrossRef]
20. Deshmukha, P.R.; Gaikwadb, V.L.; Tamanea, P.K.; Mahadikc, K.R.; Purohit, R.N. Development of stability-indicating HPLC method and accelerated stability studies for osmotic and pulsatile tablet formulations of Clopidogrel Bisulfate. *J. Pharm. Biomed. Anal.* **2019**, *165*, 346–356. [CrossRef]
21. Lee, T.-K.; Ryoo, S.-J.; Lee, Y.-S. A new method for the preparation of 2-chlorotrityl resin and its application to solid-phase peptide synthesis. *Tetrahedron Lett.* **2007**, *48*, 389–391. [CrossRef]
22. Nandi, I.; Bari, M.; Joshi, H. Study of isopropyl myristate micremulsion systems containing cyclodextrins to improve the solubility of two model hydrophobic drugs. *AAPS PharmaSciTech.* **2003**, *4*, 1–9. [CrossRef]

Sample Availability: Samples of the compounds are available from the authors.

© 2019 by the authors. Licensee MDPI, Basel, Switzerland. This article is an open access article distributed under the terms and conditions of the Creative Commons Attribution (CC BY) license (http://creativecommons.org/licenses/by/4.0/).

Article

Analysis of the Overlapped Electrochemical Signals of Hydrochlorothiazide and Pyridoxine on the Ethylenediamine-Modified Glassy Carbon Electrode by Use of Chemometrics Methods

Yue Zhang [1,†], Yan Zhou [2,†], Shujun Chen [3], Yashi You [3], Ping Qiu [3,*] and Yongnian Ni [3]

1. Port of Caofeidian Group Co., Ltd., Tangshan 063200, China
2. College of Medicine, Nanchang University, Nanchang 330031, China
3. Department of Chemistry, Nanchang University, Nanchang 330031, China
* Correspondence: pingqiu@ncu.edu.cn; Tel.: +86-791-83969514
† These authors contributed equally.

Academic Editors: Clinio Locatelli, Marcello Locatelli and Dora Melucci
Received: 1 June 2019; Accepted: 9 July 2019; Published: 11 July 2019

Abstract: In this work, the electrochemical behavior of hydrochlorothiazide and pyridoxine on the ethylenediamine-modified glassy carbon electrode were investigated by differential pulse voltammetry. In pH 3.4 Britton-Robinson (B-R) buffer solution, both hydrochlorothiazide and pyridoxine had a pair of sensitive irreversible oxidation peaks, that overlapped in the 1.10 V to 1.20 V potential range. Under the optimum experimental conditions, the peak current was linearly related to hydrochlorothiazide and pyridoxine in the concentration range of 0.10–2.0 µg/mL and 0.02–0.40 µg/mL, respectively. Chemometrics methods, including classical least squares (CLS), principal component regression (PCR) and partial least squares (PLS), were introduced to resolve the overlapped signals and determine the two components in mixtures, which avoided the troublesome steps of separation and purification. Finally, the simultaneous determination of the two components in commercial pharmaceuticals was performed with satisfactory results.

Keywords: differential pulse voltammetry; hydrochlorothiazide; pyridoxine; chemometrics

1. Introduction

Hypertension is an independent disease characterized by blood pressure above the normal range, and its etiology is not yet clear. According to the World Health Organization (WHO), hypertension is defined as adult blood pressure exceeding 21.3/12.6 kPa (160/95 mmHg) [1,2]. As a common cardiovascular disease, the risk of hypertension is not only revealed in high blood pressure, but also in a variety of pathophysiological changes, including changes in cardiovascular structure and function, nephropathy, encephalopathy and retinopathy, causing coronary atherosclerosis, cerebrovascular sclerosis and ultimately life-threatening [3–5]. There are an estimated one billion cases of hypertension in the world, which is predicted to rise to over 1.5 billion by 2015 [6,7]. In China economic development has changed people's habits and diet, the elderly population has grown and the incidence of hypertension has risen to 27.8% [2,8]. Clinical studies have found that reasonable drug treatment can effectively control blood pressure and various symptoms [9,10]. Moreover, it can reduce the incidence of some related diseases, thereby improving people's quality of life [11].

Among hypertensive patients, it is found that a single antihypertensive drug cannot lower blood pressure to normal, so two or more antihypertensive drugs are necessary to make blood pressure normal [12–14]. Each of the drug components of compound medicines have synergistic effect, which not only enhances the antihypertensive effect, but also counteracts the adverse side effects [15–17]. Most compound antihypertensives contain the diuretic-antihypertensive drug hydrochlorothiazide and a small amount of vitamin B_6 (pyridoxine) to protect the heart and prolong the antihypertensive effect.

Hitherto, various analytical methods have been reported for the separation and determination of the components in compound preparations. It has been reported that HPLC can be used in combination with various detection techniques, such as electrochemical detection [18], mass spectrometry [19,20], desorption electrospray ionization [21], size-exclusion chromatography [22], chemiluminescence [23], or spectrophotometry [24]. The electroanalytical chemistry method, with its advantages of higher sensitivity and selectivity, is preferable, and it has become a common method for drug analysis [25]. In electrochemistry, there are sereval ways to avoid overlapping signals: (1) selection of an appropriate pH and electrolyte; (2) analyte derivation; and (3) application of chemometrics. Among them, statistical chemometrics methods are environmentally friendly, and can avoid the troublesome steps of separation and purification. However, complex background currents often exist in voltammetric and polarographic analysis of drugs. The oxidation or reduction processes of electrodes are often irreversible, and their peak potentials will change with the change of drug concentration. This non-linear additive system will have a serious impact on the determination [26–28]. Moreover, it is extremely difficult to directly measure the content of a single component with serious overlap in a complex mixture. Therefore, the application of chemometrics in pharmaceutical electroanalytical chemistry has attracted extensive attention [29]. Classical least squares (CLS), principal component regression (PCR) and partial least squares (PLS) are commonly used [30].

In this work, the differential pulse voltammetry technique was used to investigate the electrochemical behavior of hydrochlorothiazide and pyridoxine on an ethylenediamine-modified glassy carbon electrode. The fabricated electrode has good stability and reproducibility [31,32], and the modified version has better electrochemical charateristics [33]. Their overlapping peaks were resolved by chemometrics, which avoided the tedious step of separation and purification, simplified the determination process and realized the simultaneous determination of two components in commercial pharmaceuticals.

2. Results and Discussion

2.1. Voltammetry at a Modified Electrode

Figure 1 shows the differential pulse voltammograms of 0.4 µg/mL hydrochlorothiazide and 0.2 µg/mL pyridoxine at the ethylenediamine-modified glassy carbon electrode (Figure 1a) and bare glassy carbon electrode (Figure 1b), respectively, in pH 3.4 B-R buffer.

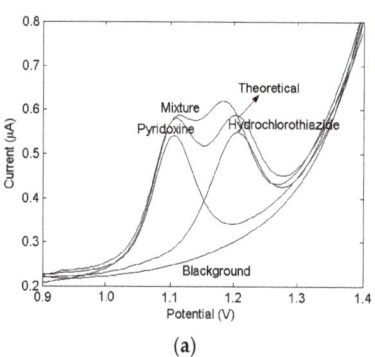
(a)

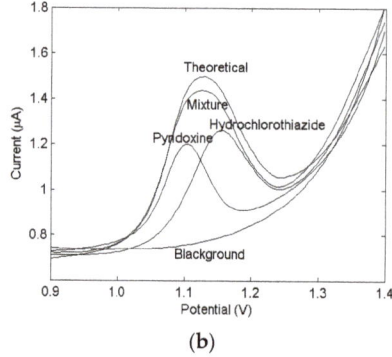
(b)

Figure 1. The differential pulse voltammograms (DPV) of 0.4 µg/mL hydrochlorothiazide and 0.2 µg/mL at the ethylenediamine-modified glassy carbon electrode (**a**) and bare glassy carbon electrode (**b**) respectively in the pH 3.4 B-R buffer.

The voltammetric curves of each compound and their mixtures show the maximum peak potentials for hydrochlorothiazide and pyridoxine at 1.11V and 1.22V, respectively, as well as the heavily overlapped nature of the composite voltammograms of the mixtures. This indicated that the sensitivity of the two drugs on the modified electrode was stronger than that on the bare electrode, and the peak potential difference between the two drugs is larger, which was helpful for chemometric analysis.

Ethylenediamine modifies the surface of the glassy carbon electrode by covalent bonding, which is based on oxidation of amino groups [34,35]. Primary and secondary amines can form amine radical cations through the oxidation of the amino group in anhydrous ethanol or acetonitrile solution. The radical is further bonded on the surface of glassy carbon or carbon fibers to form C-H covalent bonds [36,37]. Therefore, the modification method is also called the "amine radical cation method".

It is generally believed that the modifying cation and the electrode are vertically oriented. Primary amines are easily grafted on the surface of electrode during the electrooxidation, while secondary amines oxidize to form a monolayer with low coverage, and tertiary amines cannot form a monolayer. Therefore, it can be concluded that the steric hindrance of the substituent affects the active site of radicals on the surface of the electrode during the amine oxidation process. The bonding reaction requires protons to be detached from the amine cation radical, so this reaction path is not feasible for tertiary amines [38]. Ethylenediamine is a typical aliphatic diamine, which exhibits characteristics of both primary and secondary amines. The chemical properties of ethylenediamine are related to the number of hydrogen atoms substituents on the functional group (NH_2). In this experiment, it seems the special chemical structure coordinates on the surface of the electrode to form an ethylenediamine-modified electrode.

As for the repeatability of the modified electrode, the results showed that the relative standard deviation (n = 3) was less than 3.8%. Regarding stability, samples containing the drugs were stored in the refrigerator at 4 °C. After 15 days, the current was measured with a standard deviation of less than 3.4%. The results proved that the modified electrode was stable and the processes have good repeatability.

2.2. Selection of Buffer Solution

This experiment investigated the effects of various buffer solutions on the electrochemical response of drugs. The tartaric acid-sodium tartrate buffer, acetic acid-sodium acetate buffer, disodium hydrogen phosphate-citric acid buffer and Britton-Robinson buffer were investigated. It was found that the sensitivity and symmetry of the oxidation peaks of each component in the Britton-Robinson buffer were highest, so the B-R buffer was selected.

The effect of acidity on the voltammetric curves of the drugs was also investigated. The electrochemical voltammetric curves of hydrochlorothiazide and pyridoxine in a series of B-R buffer solutions at pH 1.98–11.98 were measured, respectively. It was found that the peak current of the drugs increased and then decreased with the increase of pH value, and the peak shape gradually broadened. The peak potential of the drugs moved negatively with the increase of pH and showed a linear relationship. The linear equations were expressed as below:

$$\text{Hydrochlorothiazide: } E_p = -0.0468\, pH + 1.3202\ (R = 0.999), \tag{1}$$

$$\text{Pyridoxine: } E_p = -0.0515\, pH + 1.2710\ (R = 0.998), \tag{2}$$

In acidic B-R buffer (pH 3.4), the drugs had better peak shape and higher peak sensitivity. Thus, these results clearly indicated that the pH 3.4 B-R buffer solution should be selected as supporting electrolyte in this experiment.

2.3. Cyclic Voltammetry to Study the Adsorptivity of the Electrode Reaction

The effect of the scan rate and the square root of the scan rate on peak current by cyclic voltammetry for the different drugs is examined (Figure 2). The peak current of both components showed a linear relationship with the change of scan rate, but had a upward bending curve with square root of scan rate, illustrating that the electrode process is controlled by the adsorption rate at that time.

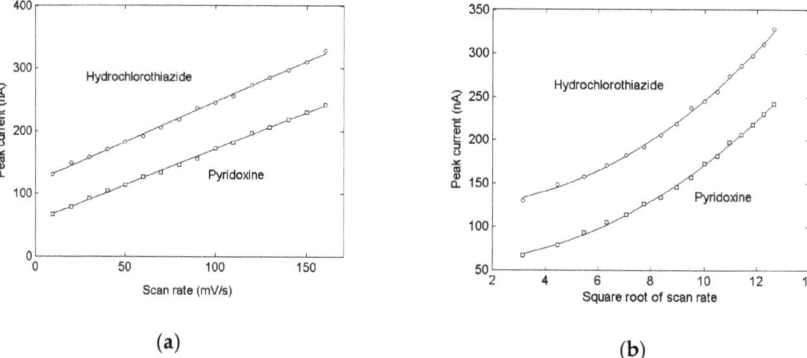

Figure 2. Effect of scan rate (a) and square root of scan rate (b) on peak current of the drugs.

2.4. *Linearity Ranges and Limits of Detection for Drugs*

Under the optimal experimental conditions, the drug concentrations had linear relationships with the electrochemical signals. Figure 3 shows voltammetric curves for the determination of hydrochlorothiazide (a) and pyridoxine (b). There is a good linear relationship between the peak current and the drugs concentration in the range of 0.02–0.40 µg/mL for pyridoxine and 0.10–2.0 µg/mL for hydrochlorothiaxide. The regression parameters are summarized in Table 1, and the accuracy of the determination of these drugs has also been established by analysing the minimum concentration calibration graph for six known solutions, i.e., 0.02 µg/mL (see Table 1).

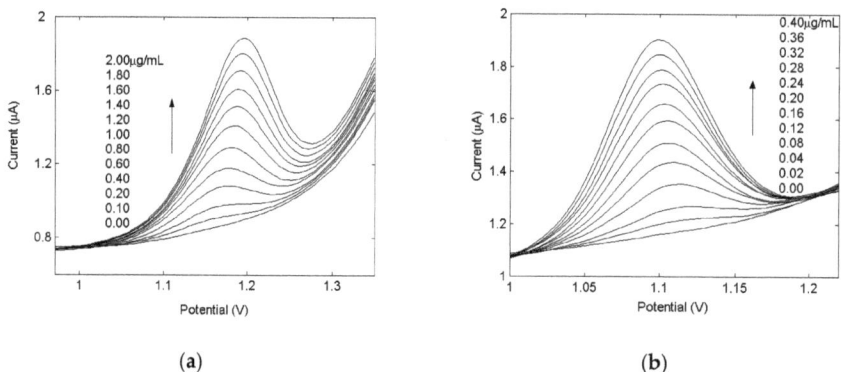

Figure 3. Voltammogram of hydrochlorothiazide (a) and pyridoxine (b) for different concentrations.

Table 1. The parameters of linear equation obtained by the proposed method.

Parameters	Pyridoxine	Hydrochlorothiazide
Sample number (n)	11	11
Linear range (µg/mL)	0.02–0.40	0.10–2.0
Intercept (nA)	14.38	12.12
Slope(nA·mL/µg)	0.60	0.20
Correlation coefficient	0.9995	0.9995
Limit of detection (ng/mL)	7.00	37.60

The relative standard deviation (R.S.D.) values obtained were 1.7% and 2.1% for pyridoxine and hydrochlorothiazide, respectively, and the detection limit values were 7.00 and 37.60 ng/mL for pyridoxine and hydrochlorothiazide, respectively. Therefore, it can be concluded that the proposed electrochemical analysis method for the determination of individual drugs is reliable.

2.5. Prediction of Synthetic Mixtures of Drug Compounds

Before the determination of unknown mixtures by chemometrics, a set of mathematical models should be developped to correct the concentration of each component in the mixtures. According to the four-level orthogonal array design represented by $OA_{16}(4^5)$, a set of standard samples was prepared, which indicated that a data set of 16 samples was required. The calibration set is based on the given concentration range for 0.02–1.0 µg/mL, as it is usually desirable to test the lower performance of the calibration rather than the higher concentrations [39]. The prediction power of the calibration model was then assessed using another set of samples consisting of 12 synthetic mixtures. In our work, CLS, PLS, and PCR models were established and their prediction errors were compared. CLS method is often called K matrix method, which is a commonly used multivariate correction method. This method is based on multiple linear regression and is frequently used for quantitative voltammetric analysis, which was a much common multivariate calibration method [40]. PCR and PLS, which are powerful multivariate statistical tools and are available as commercial software for laboratory computers [41], are based on factor analysis [42]. These methods have many of the full-soectrum advantages of the CLS method and have also been successfully applied for voltammetric analysis [43].

From the measured results in Table 2, the best prediction results could be obtained by the PLS method, while the CLS method was the worst. Because of the background current and the interaction between components in voltammetric analysis, the non-linear superposition of voltammograms would occur and the CLS method had some difficulties in analyzing such systems, while PCR and PLS are calibration methods based on factor analysis, which could better solve this kind of non-linear problem [44]. In the PCR and PLS methods, a certain number of factors must be selected to perform a matrix factorization, so that the response matrix could be reproduced within the experimental error range. The selection of factors has a great impact on the calculation results. Generally, when the number of factors was less than or slightly larger than the component number, the relative standard deviation of the calculation results was smaller [45]. Figure 4 shows the relationship between the relative error and the number of factors. In this way, when the PCR method was used, the number of factor was 4, the relative error was smaller and the relative error is the minimum when the number of factor was 4 by PLS method.

Table 2. Prediction results for pyridoxine(PX) and hydrochlorothiazide(HC) validation samples by different chemometrics methods (µg/mL).

Sample	Added		Found (CLS)		Found (PLS) [1]		Found (PCR) [2]	
	PX	HC	PX	HC	PX	HC	PX	HC
1	0.025	0.100	0.212	0.322	0.025	0.173	0.024	0.143
2	0.025	0.275	0.194	0.418	0.026	0.260	0.027	0.187
3	0.050	0.600	0.150	0.641	0.041	0.572	0.043	0.506
4	0.050	0.900	0.111	0.834	0.049	0.891	0.049	0.840
5	0.110	0.100	0.239	0.195	0.133	0.068	0.134	0.057
6	0.110	0.275	0.210	0.372	0.105	0.272	0.106	0.251
7	0.110	0.600	0.173	0.558	0.113	0.563	0.114	0.542
8	0.110	0.900	0.141	0.759	0.101	0.888	0.101	0.843
9	0.170	0.100	0.255	0.162	0.188	0.811	0.189	0.096
10	0.170	0.275	0.228	0.303	0.173	0.264	0.174	0.278
11	0.170	0.600	0.194	0.508	0.164	0.568	0.165	0.552
12	0.170	0.900	0.156	0.680	0.162	0.873	0.162	0.858

Table 2. *Cont.*

Sample	Added		Found (CLS)		Found (PLS) [1]		Found (PCR) [2]	
	PX	HC	PX	HC	PX	HC	PX	HC
RPE$_S$ (%) [3]			51.8	23.2	8.18	5.47	8.41	9.33
Recovery (%) [4]			277.4	134.4	100.1	99.19	100.5	91.58
RPE$_T$ (%) [3]			46.8		7.04		8.86	

[1] The factors of PLS is 4. [2] The factors of PCR is 4. [3] RPE$_S$ (%) and RPE$_T$ (%) are relative prediction errors for single and total components, respectively. [4] Recovery (%) = $100 \times \sum (c_{ij,pred} - c_{ij,added})/n$, where n is the number of samples, c_{ij} is the concentration of the *j*-th component in the *i*-th sample.

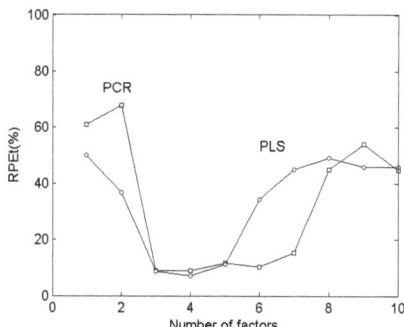

Figure 4. The relationship of PRE$_t$ and the number of factors used for PLS and PCR.

3. Detection of the Drugs in Real Samples

The proposed method was applied for the determination of the drugs in four commercial tablets, produced by two different pharmaceutical companies. In each case, the sugar coating was removed from ten tablets which were then ground into a powder and dissolved in doubly distilled water. After filtering three times the filtrate was collected and diluted to 50 mL with distilled water.

According to the described PLS procedure, the samples were analyzed with the results listed in Table 3. The values measured by the proposed method were consistent with the target values. The average relative error between two methods was 3.1%, indicating the method provided excellent precision in detecting the drugs. The recovery rate was between 92.5% and 106.2%, with RSD (n = 5) less than 2.9%. It could be seen that the simultaneous determination of hydrochlorothiazide and pyridoxine in compound medicines could be achieved by the proposed electrochemical method without labelling, and the results were satisfactory.

Table 3. Pyridoxine(PX) and hydrochlorothiazide(HC) in commercial tablets as determined by the proposed method (mg/tablet).

Sample	Target Values		Found by This Method		Recovery(%)	
	PX	HC	PX	HC	PX	HC
Tablet 1[1]	0.5	1.6	0.47 ± 0.03	1.48 ± 0.02	94.0	92.5
Tablet 2[2]	0.5	1.6	0.44 ± 0.04	1.69 ± 0.03	88.0	105.6
Tablet 3[3]	1.0	3.1	1.01 ± 0.02	3.29 ± 0.01	101.0	106.1
Tablet 4[4]	1.0	3.1	1.04 ± 0.02	3.25 ± 0.03	104.0	104.8

[1] Jinhua Yexing Pharmaceutical Co., Ltd., Shanxi. Lot code. 061108. [2] Dongzhitang Pharmaceutical Co., Ltd., Anhui. Lot code. 06101102. [3] Yinhe Pharmaceutical Factory, Jilin. Lot code. 20050801. [4] Fenhe Pharmaceutical Co., Ltd., Shanxi. Lot code. 0701031.

4. Materials and Methods

4.1. Reagents and Apparatus

Hydrochlorothiazide (0.1 mg/mL) and pyridoxine (0.02 mg/mL, ethanol) were obtained from Shanghai Biochemical Reagent Co. Ltd. (Shanghai, China). Ethylenediamine (>99%) was dissolved to 1 mol/L solution with anhydrous ethanol for use as the ethylenediamine-ethanol modifier. A Britton-Robinson (B-R) buffer solution of pH 3.4 was prepared by adding 15 mL 0.2 mol/L sodium hydroxide into 100 mL of a mixed acid containing 0.04 mol/mL each of orthophosphoric acid, acetic acid, and boric acid.

All chemicals used (Sinopharm Chemical Reagent Co., Ltd. Shanghai, China) were of analytical-reagent grade and all the solutions were prepared with doubly distilled water. The experiments were carried out at room temperature.

A CHI 660A electrochemical workstation (Shanghai CH Instruments Co., Shanghai China) equipped with a BAS C-1 cell stand was used for voltammetric measurements. The three electrode system consisted of a glassy carbon electrode as work electrode, Ag/AgCl reference electrode and platinum wire as auxiliary electrode. The pH measurements were performed on a SA720 meter (Thermo Orion, Waltham, MA, USA).

4.2. Procedure

4.2.1. Modification of Glassy Carbon Electrode

Firstly the glassy carbon electrode was polished on 1200-grit Carbimet metallographic sandpaper, and to a mirror finish with 1.0 μm α-alumina powder, 0.3 μm α-alumina powder and 0.05 μm γ-alumina powder in turn. Then HNO_3 (1:1), anhydrous ethanol and doubly distilled water were used to clean the electrode with ultrasonic irradiation, respectively. Finally, the activated electrode was dipped in 1 mol/L ethylenediamine-ethanol solution for 12 h and the ethylene- diamine-modified glassy carbon electrode was ready for use.

4.2.2. Procedure

A suitable amount of standard solution of hydrochlorothiazide and pyridoxine or their mixed solution were prepared in the electrolytic cell, together with B-R buffer (2 mL pH 3.4). The solution was diluted to 10 mL with doubly distilled water and shaken evenly. Then the ethylenediamine-modified glassy carbon electrode was placed, the solution was thoroughly mixed by stirring for 15 s. After a 10 s static period, a differential pulse voltammetric scan was run from 0.9 to 1.4 V. The resulting voltammograms were sampled by a computer at 4 mV intervals. The experiment was completed at 20 °C room temperature.

5. Conclusions

In this work, the use of a modified electrode to improve the selectivity of voltammetric data in complex matrices is the issue, even if the separation is not complete and requires the use of chemometric tools. Furthermore, PCL and PCR seem to perform adequately for deconvolution of the voltammetric signals. Chemometrics methods can thus be used for the quantitative analysis of two components in commercial drugs with simple operation and satisfactory results.

Author Contributions: Y.Z. (Yue Zhang) substantially contributed the conception of the work; Y.Z. (Yan Zhou) drafted the work and contributed with Y.Z. (Yue Zhang) equally; S.C. analyzed the data; Y.Y. interpreted of data; P.Q. substantially contributed the design of the work and finally approved the published version; Y.N. agreed to be accountable for all aspects of the work.

Funding: This research was funded by the National Natural Science Foundation of China (21765015, 21808099) and the Jiangxi Province Food and Drug Administration Science Foundation (2016SP04), China.

Conflicts of Interest: The authors declare no conflict of interest. Sources of support is not in study. All the authors listed have approved the manuscipt. The decision had no such invedement.

References

1. Mills, K.T.; Bundy, J.D.; Kelly, T.N.; Reed, J.E.; Kearney, P.M.; Reynolds, K.; Chen, J.; He, J. Global burden of hypertension: Analysis of population-based studies from 89 countries. *J. Hypertens.* **2015**, *33*, E2. [CrossRef]
2. Mendis, S.; Lindholm, L.H.; Mancia, G.; Whitworth, J.; Alderman, M.; Lim, S.; Heagerty, T. World health organization (who) and international society of hypertension (ish) risk prediction charts: Assessment of cardiovascular risk for prevention and control of cardiovascular disease in low and middle-income countries. *J. Hypertens.* **2007**, *25*, 1578–1582. [CrossRef] [PubMed]
3. Tientcheu, D.; Ayers, C.; Das, S.R.; McGuire, D.K.; de Lemos, J.A.; Khera, A.; Kaplan, N.; Victor, R.; Vongpatanasin, W. Target Organ Complications and Cardiovascular Events Associated With Masked Hypertension and White-Coat Hypertension: Analysis From the Dallas Heart Study. *J. Am. Soc. Hypertens.* **2015**, *66*, 2159–2169. [CrossRef]
4. Farkas, Z.C.; Chugh, P.; Frager, S.; Haq, K.F.; Khan, M.A.; Solanki, S.; Esses, E.; Veillette, G.; Bodin, R. Periampullary Variceal Bleeding: An Atypical Complication of Portal Hypertension. *Case Rep. Gastrointest. Med.* **2018**, *2018*, 4643695. [CrossRef] [PubMed]
5. Chang, Y.; Choi, G.S.; Lim, S.M.; Kim, Y.J.; Song, T.J. Interarm Systolic and Diastolic Blood Pressure Difference Is Diversely Associated With Cerebral Atherosclerosis in Noncardioembolic Stroke Patients. *Am. J. Hypertens.* **2017**, *31*, 35–42. [CrossRef] [PubMed]
6. Sarkar, C.; Webster, C.; Gallacher, J. Neighbourhood walkability and incidence of hypertension: Findings from the study of 429,334 UK Biobank participants. *Int. J. Hyg. Envir. Heal.* **2018**, *221*, 458–468. [CrossRef] [PubMed]
7. Neutel, J. State of Hypertension Control. In *Combination Therapy in Hypertension*; Springer Healthcare Ltd.: Orange, CA, USA, 2011; pp. 1–11. [CrossRef]
8. Li, Y.; Yang, L.; Wang, L.; Zhang, M.; Huang, Z.; Deng, Q.; Zhou, M.; Chen, Z.; Wang, L. Burden of hypertension in China: A nationally representative survey of 174,621 adults. *Int. J. Cardiol.* **2017**, *227*, 516–523. [CrossRef] [PubMed]
9. Bosch, J.; Berzigotti, A.; Garcia-Pagan, J.C.; Abraldes, J.G. The management of portal hypertension: Rational basis, available treatments and future options☆. *J. Hepatol.* **2008**, *48*, S68–S92. [CrossRef]
10. Tavassoli, Z.; Taghdir, M.; Ranjbar, B. Renin inhibition by soyasaponin I: A potent native anti-hypertensive compound. *J. Biomol. Struct. Dyn.* **2018**, *36*, 166–176. [CrossRef]
11. Oparil, S.; Schmieder, R.E. New approaches in the treatment of hypertension. *Circ. Res.* **2015**, *116*, 1074–1095. [CrossRef]
12. Wenyi, F.; Yiqun, W.U.; Yang, C.; Yang, C.; Xueying, Q.; Xun, T.; Dafang, C.; Siyan, Z.; Yonghua, H. The efficacy and safety of compound anti-hypertensive tablets (NO.O) in patients with elderly isolated systolic hypertension. *Mod. Prev. Med.* **2012**, *39*, 1008–1010.
13. Ma, L.; Han, R.; Li, L.; Li, Z.; Sun, F.; Diao, L.; Tang, Z. Trends in the prevalence of antihypertensive drug treatment in the Beijing Longitudinal Study of Aging. *Arch. Gerontol. Geriat.* **2018**, *74*, 44–48. [CrossRef] [PubMed]
14. Huang, D.; Huang, J. Clinical Application and Research Progress of the Compound Antihypertensive. *Drug Eval.* **2011**, *08*, 8–13. [CrossRef]
15. Li, B.; Yang, Z.B.; Lei, S.S.; Su, J.; Jin, Z.W.; Chen, S.H.; Lv, G.Y. Combined Antihypertensive Effect of Paeoniflorin Enriched Extract and Metoprolol in Spontaneously Hypertensive Rats. *Pharmacogn. Mag.* **2018**, *14*, 44–52. [CrossRef] [PubMed]
16. Zhu, M.; Gao, J.L.; Jin, G.Q. Effects and Expense of Blood Pressure Control on Hypertensive Patients Using Different Treatment Methods: A Comparative Study. *Chin. Gen. Pract.* **2016**, *19*, 96–99. [CrossRef]
17. Lu, X.M. Simultaneous Determination of Content and Uniformity of Hydrochlorothiazide and Promethazine Hydro- chloride in Compound Kendir Leaves Tablets I by HPLC. *Chin. Pharm.* **2014**, *17*, 1891–1893.
18. Long, Z.; Gamache, P.; Guo, Z.; Pan, Y.; Zhang, Y.; Liu, X.; Jin, Y.; Liu, L.; Liang, L.; Li, R. A highly sensitive high performance liquid chromatography-electrochemical detection method for the determination of five phenolic compounds from salvia miltiorrhiza. *Se pu Chin. J. Chromatogr.* **2017**, *35*, 897–905. [CrossRef]

19. Sanford, A.A.; Isenberg, S.L.; Carter, M.D.; Mojica, M.A.; Mathews, T.P.; Harden, L.A.; Takeoka, G.R.; Thomas, J.D.; Pirkle, J.L.; Johnson, R.C. Quantitative HPLC-MS/MS analysis of toxins in soapberry seeds: Methylenecyclopropylglycine and hypoglycin A. *Food Chem.* **2018**, *264*, S0308814618307349. [CrossRef]
20. Wilson, I.D.; Plumb, R.; Granger, J.; Major, H.; Williams, R.; Lenz, E.M. Hplc-ms-based methods for the study of metabonomics. *J. Chromatogr. B Analyt. Technol. Biomed. Life Sci.* **2005**, *817*, 67–76. [CrossRef]
21. Kotrebai, M.; Birringer, M.; Tyson, J.F.; Block, E.; Uden, P.C. Selenium speciation in enriched and natural samples by hplc-icp-ms and hplc-esi-ms with perfluorinated carboxylic acid ion-pairing agents. *Analyst* **2000**, *125*, 71–78. [CrossRef]
22. Her, N.; Amy, G.; Mcknight, D.; Sohn, J.; Yoon, Y. Characterization of dom as a function of mw by fluorescence eem and hplc-sec using uva, doc, and fluorescence detection. *Water Res.* **2003**, *37*, 4295–4303. [CrossRef]
23. Fan, S.; Zhang, L.; Lin, J. Post-column detection of benzenediols and 1,2,4-benzenetriol based on acidic potassium permanganate chemiluminescence. *Talanta* **2006**, *68*, 646–652. [CrossRef] [PubMed]
24. Gil-Alegre, M.E.; Barone, M.L.; Torres-Suárez, A.L. Extraction and determination by liquid chromatography and spectrophotometry of naloxone in microparticles for drug-addiction treatment. *J. Sep. Sci.* **2005**, *28*, 2086–2093. [CrossRef] [PubMed]
25. Gao, P.; Ji, M.; Fang, X.; Liu, Y.; Yu, Z.; Cao, Y.; Sun, A.; Zhao, L.; Zhang, Y. Capillary electrophoresis—Mass spectrometry metabolomics analysis revealed enrichment of hypotaurine in rat glioma tissues. *Anal. Biochem.* **2017**, *537*. [CrossRef] [PubMed]
26. Bonfiglio, R.; King, R.C.; Olah, T.V.; Merkle, K. The effects of sample preparation methods on the variability of the electrospray ionization response for model drug compounds. *Rapid Commun. Mass Spectrom.* **1999**, *13*, 1175–1185. [CrossRef]
27. Erdem, A.; Karadeniz, H.; Caliskan, A. Dendrimer modified graphite sensors for detection of anticancer drug Daunorubicin by voltammetry and electrochemical impedance spectroscopy. *Analyst* **2011**, *136*, 1041–1045. [CrossRef]
28. Ferris, M.J.; Calipari, E.S.; Yorgason, J.L.; Jones, S.R. Examining the Complex Regulation and Drug-InducedPlasticity of Dopamine Release and Uptake Using Voltammetry in BrainSlices. *ACS Chem. Neurosci.* **2013**, *4*, 693–703. [CrossRef]
29. Norouzi, P.; Haji-Hashemi, H.; Larijani, B.; Aghazadeh, M.; Pourbasheer, E.; Ganjali, M.R. Application of New Advanced Electrochemical Methods Combine with Nano-Based Materials Sensor in Drugs Analysis. *Curr. Anal. Chem.* **2016**, *13*, 70–80. [CrossRef]
30. Dinc, E. Spectral analysis of benazepril hydrochloride and hydrochlorothiazide in pharmaceutical formulations by three chemometric techniques. *Anal. Lett.* **2002**, *35*, 1021–1039. [CrossRef]
31. Zhu, Y.H.; Zhang, Z.L.; Pang, D.W. Electrochemical oxidation of theophylline at multi-wall carbon nanotube modified glassy carbon electrodes. *J. Electroanal. Chem.* **2005**, *581*, 303–309. [CrossRef]
32. Sakthinathan, S.; Kokulnathan, T.; Chen, S.M.; Karthik, R.; Chiu, T.W. Ecofriendly preparation of graphene sheet decorated with ethylenediamine copper (II) complex composite fabricated electrode for selective detection of hydroquinone in water. *Inorg. Chem. Front.* **2017**, *5*, 490–500. [CrossRef]
33. Zhou, J.; Wang, L.; Chen, Q.; Wang, Y.; Fu, Y. Selective response of antigen-antibody reactions on chiral surfaces modified with 1,2-diphenylethylenediamine enantiomers. *Surf. Interface Anal.* **2012**, *44*, 170–174. [CrossRef]
34. Yang, G.; Liu, B.; Dong, S. Covalent modification of glassy carbon electrode during electrochemical oxidation process of 4-aminobenzylphosphonic acid in aqueous solution. *J. Electroanal. Chem.* **2005**, *585*, 301–305. [CrossRef]
35. Sumalekshmy, S.; Gopidas, K.R. Reaction of aromatic amines with Cu(ClO4)2 in acetonitrile as a facile route to amine radical cation generation. *Chem. Phys. Lett.* **2005**, *413*, 294–299. [CrossRef]
36. Morris, S.A.; Wang, J.; Zheng, N. The Prowess of Photogenerated Amine Radical Cations in Cascade Reactions: From Carbocycles to Heterocycles. *Acc. Chem. Res.* **2016**, *49*, 1957–1968. [CrossRef] [PubMed]
37. Hu, J.; Wang, J.; Nguyen, T.H.; Zheng, N. The chemistry of amine radical cations produced by visible light photoredox catalysis. *Beilstein J. Org. Chem.* **2013**, *9*, 1977–2001. [CrossRef] [PubMed]
38. Huang, B.; Jia, N.; Chen, L.; Tan, L.; Yao, S. Electrochemical impedance spectroscopy study on polymerization of\r, l\r, -lysine on electrode surface and its application for immobilization and detection of suspension cells. *Anal. Chem.* **2014**, *86*, 6940–6947. [CrossRef] [PubMed]
39. Tabaraki, R.; Khodabakhshi, M. Multidye Biosorption: Wavelet Neural Network Modeling and Taguchi L 16 Orthogonal Array Design. *Clean Soil Air Water* **2017**, *45*, 1500499. [CrossRef]

40. Ni, Y.; Wang, L.; Kokot, S. Simultaneous determination of nitrobenzene and nitro-substituted phenols by differential pulse voltammetry and chemometrics. *Anal. Chim. Acta* **2001**, *431*, 101–113. [CrossRef]
41. Ghasemi, J.; Niazi, A. Simultaneous determination of cobalt and nickel. comparison of prediction ability of pcr and pls using original, first and second derivative spectra. *Microchem. J.* **2001**, *68*, 1–11. [CrossRef]
42. Crouch, S.R.; Coello, J.; Maspoch, S.; Porcel, M. Evaluation of classical and three-way multivariate calibration procedures in kinetic-spectrophotometric analysis. *Anal. Chim. Acta* **2000**, *424*, 115–126. [CrossRef]
43. Hegazy, M.A.; Elghobashy, M.R.; Yehia, A.M.; Mostafa, A.A. Simultaneous determination of metformin hydrochloride and pioglitazone hydrochloride in binary mixture and in their ternary mixture with pioglitazone acid degradate using spectrophotometric and chemometric methods. *Drug Test. Anal.* **2010**, *1*, 339–349. [CrossRef] [PubMed]
44. Henseler, J.; Hubona, G.; Ray, P. Using PLS Path Modeling in New Technology Research: Updated Guidelines. *Ind. Manage. Data Syst.* **2016**, *116*, 2–20. [CrossRef]
45. Zhang, F.Y.; Zhang, R.Q.; Ge, J.; Chen, W.C.; Yang, W.Y.; Du, Y.P. Calibration transfer based on the weight matrix (CTWM) of PLS for near infrared (NIR) spectral analysis. *Anal. Methods* **2018**, *10*, 2169–2179. [CrossRef]

Sample Availability: Samples of the hydrochlorothiazide and pyridoxine are available from the authors.

© 2019 by the authors. Licensee MDPI, Basel, Switzerland. This article is an open access article distributed under the terms and conditions of the Creative Commons Attribution (CC BY) license (http://creativecommons.org/licenses/by/4.0/).

Article

Novel MIPs-Parabens based SPE Stationary Phases Characterization and Application

Angela Tartaglia [1], Abuzar Kabir [2,*], Songul Ulusoy [3], Halil Ibrahim Ulusoy [4], Giuseppe Maria Merone [5], Fabio Savini [6], Cristian D'Ovidio [7], Ugo de Grazia [8], Serena Gabrielli [9], Fabio Maroni [1], Pantaleone Bruni [1], Fausto Croce [1], Dora Melucci [10], Kenneth G. Furton [2] and Marcello Locatelli [1,*]

[1] Department of Pharmacy, University of Chieti–Pescara "G. d'Annunzio", Via dei Vestini 31, 66100 Chieti, Italy; angela.tartaglia@unich.it (A.T.); fabio.maroni@unich.it (F.M.); pantaleonebruni@libero.it (P.B.); fausto.croce@unich.it (F.C.)
[2] Department of Chemistry and Biochemistry, International Forensic Research Institute, Florida International University, 11200 SW 8th St, Miami, FL 33199, USA; furtonk@fiu.edu
[3] Department of Chemistry, Faculty of Science, Cumhuriyet University, Sivas 58140, Turkey; sonulusoy@yahoo.com
[4] Department of Analytical Chemistry, Faculty of Pharmacy, Cumhuriyet University, Sivas 58140, Turkey; hiulusoy@yahoo.com
[5] Department of Neuroscience, Imaging and Clinical Sciences, University of Chieti–Pescara "G. d'Annunzio", Via dei Vestini 31, 66100 Chieti, Italy; giuseppe.merone@unich.it
[6] Pharmatoxicology Laboratory—Hospital "Santo Spirito", Via Fonte Romana 8, 65124 Pescara, Italy; fabio.savini@ausl.pe.it
[7] Department of Medicine and Aging Sciences, Section of Legal Medicine, University of Chieti–Pescara "G. d'Annunzio", Via dei Vestini 31, 66100 Chieti, Italy; cristian.dovidio@unich.it
[8] IRCCS Neurological Institute Foundation Carlo Besta, Laboratory of Neurological Biochemistry and Neuropharmacology, Via Celoria 11, 20133 Milan, Italy; ugo.degrazia@istituto-besta.it
[9] Chemical Science Department, School of Science and Technology, University of Camerino, Piazza Cavour 19/f, 62032 Camerino, Italy; serena.gabrielli@unicam.it
[10] Department of Chemistry "G. Ciamician", University of Bologna, Via Selmi 2, 40126 Bologna, Italy; dora.melucci@unibo.it
* Correspondence: akabir@fiu.edu (A.K.); marcello.locatelli@unich.it (M.L.); Tel.: +1-305 348 2396 (A.K.); +39-0871-3554590 (M.L.); Fax: +1-305 348 4172 (A.K.); +39-0871-3554911 (M.L.)

Academic Editors: Victoria Samanidou and Alessandra Gentili
Received: 24 July 2019; Accepted: 10 September 2019; Published: 13 September 2019

Abstract: In this work, the synthesis, characterization, and application of novel parabens imprinted polymers as highly selective solid-phase extraction (SPE) sorbents have been reported. The imprinted polymers were created using sol–gel molecular imprinting process. All the seven parabens were considered herein in order to check the phase selectivity. By means of a validated HPLC-photodiode array detector (PDA) method all seven parabens were resolved in a single chromatographic run of 25 min. These SPE sorbents, *in-house* packed in SPE empty cartridges, were first characterized in terms of extraction capability, breakthrough volume, retention volume, hold-up volume, number of theoretical plates, and retention factor. Finally, the device was applied to a real urine sample to check the method feasibility on a very complex matrix. The new paraben imprinted SPE sorbents, not yet present in the literature, potentially encourage the development of novel molecularly imprinted polymers (MIPs) to enhance the extraction efficiency, and consequently the overall analytical performances, when the trace quantification is required.

Keywords: MIPs; parabens; biological matrix; extraction procedure; HPLC-PDA; stationary phase characterization

1. Introduction

Parabens are alkyl esters of *p*-hydroxybenzoic acid, commonly used as antimicrobial agents and preservatives in food products, pharmaceutical preparations, and cosmetic products [1]. This family includes methyl paraben (MPB), ethyl paraben (EPB), iso-propyl paraben (iPPB), propyl paraben (PPB), iso-butyl paraben (iBPB), butyl paraben (BPB), and benzyl paraben (BzPB). Their molecular formula, molecular weight, and chemical structure are shown in Table 1. Low cost, broad spectrum of activity, and chemical and thermal stability explain their widespread application compared to other alternatives [2].

Table 1. General chemical–physical characteristics of parabens and their chemical structures.

	Molecular Formula	Molecular Weight	Chemical Structure
MPB	$C_8H_8O_3$	152.15 Da	
EPB	$C_9H_{10}O_3$	166.17 Da	
iPPB	$C_{10}H_{12}O_3$	180.20 Da	
PPB	$C_{10}H_{11}O_3$	180.20 Da	
iBPB	$C_{11}H_{14}O_3$	194.23 Da	
BPB	$C_{11}H_{14}O_3$	194.23 Da	
BzPB	$C_{14}H_{12}O_3$	228.25 Da	

Humans may be exposed to parabens through inhalation, dermal contact, and ingestion. Parabens have been considered low toxicity compounds; nevertheless, there is growing evidence of their implication as endocrine disruption compounds (EDCs), i.e., products with negative effects on endocrine system due to their interaction with production, release, transport, metabolism, or elimination of hormones. Many studies describe the toxic effects of parabens such as obesity, diabetes, breast cancer and problems in reproductive system [3].

According to the European Union (EU) and the United States Food and Drug Administration (US FDA), the concentration of parabens should not exceed 0.4% for a single paraben and 0.8% for a mixture of parabens in the cosmetics product; the concentration in drugs must be less than 0.1% and 0.3% for individual paraben and for parabens mixture, respectively [4].

During the last years, new methods have been developed for the determination of parabens in different matrices. Due to the low concentrations of these compounds, the sample preparation represents the critical step for the isolation and analysis of these target molecules. Usually, common extraction techniques were used for pre-concentrating parabens, such as solid-phase extraction (SPE) [5–8] and solid-phase microextraction (SPME) [9]. For SPE, different materials are available for extraction, preconcentration, and sample cleaning. Such sorbents used in solid-phase extraction can be divided into three main groups: silica sorbents, polymeric sorbents, and activated or graphitized carbon. Analytical parameters such as selectivity, affinity, and extraction capacity are closely associated with the sorbent used and, consequently, various SPE materials have been developed to replace the classic ones and to increase the extractive selectivity. An example is represented by molecularly imprinted polymers (MIPs) used as SPE sorbent [1].

MIPs are formed by fixing a molecule (called a "template") on the polymer; the molecule is extracted afterward, leaving complementary cavities behind [10]. Molecular imprinting is obtained by polymerization of monomers in the presence of template molecules (Figure 1). The imprinting template should be stable under polymerization reaction. The link between template and monomer must be strong to form the pocket but also weak enough to be able to remove the template at the end. Usually, the monomer and template are in 4:1 molar ratio. The most frequently used monomers are methacrylic acid and 4-vinylpyridine [11].

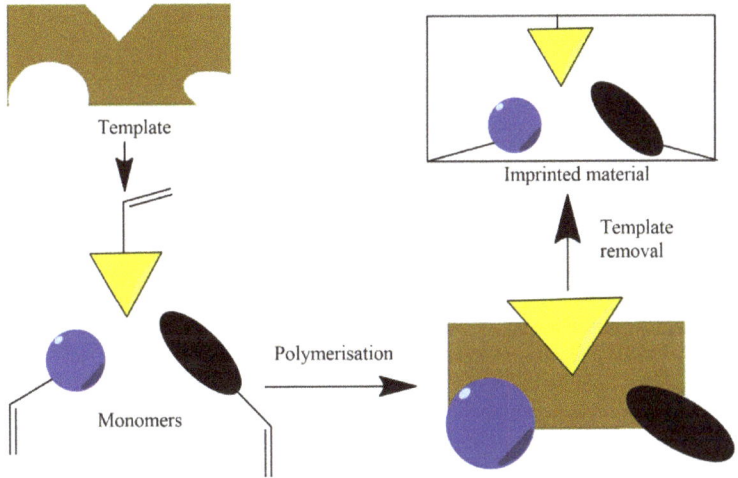

Figure 1. Synthesis of molecularly imprinted polymers (MIPs).

There are different methods used to perform molecular imprinting: covalent imprinting, non-covalent imprinting, and semi-covalent imprinting. In covalent imprinting, the template and monomer are linked by covalent bonds; after polymerization, the template is removed from the polymer through the cleavage of covalent bond. The template is replaced from target analytes that bind forming the same covalent links. In non-covalent imprinting, template and monomer are linked by non-covalent interactions such as hydrogen bonding, ionic interactions, π–π interactions, and van der Waals forces. After polymerization, the template is removed, and the target analytes bind the polymer via same non-covalent interactions. This technology is the most popular for the synthesis of MIP. In semi-covalent approach, the template is linked covalently to the monomer but is rebinding by non-covalent interactions [12].

Among several MIP synthesis approaches, sol–gel synthesis process for creating MIPs has drawn tremendous attention due to ruggedness of the polymers originated from hybrid inorganic–organic components, high thermal and solvent stability, and ability to maintain imprinted cavities even after multiple use as well as their characteristic high imprinting factor (IF) compared to organic MIP synthesis approach. The success of sol–gel based MIP synthesis primarily depends on the judicious selection of sol–gel functional precursor(s), networking precursor, and the reaction conditions including the catalysts. In the current study, 3-aminopropyl triethoxysilane and phenyltriethoxysilane were used as the sol–gel functional polymers to provide hydrogen bonding and π–π interactions toward the template molecules, respectively. Tetramethyl orthosilicate was used as the networking sol–gel precursor, whose primary role was to rigidly hold the template-sol–gel functional precursors complexes in the sol–gel 3D network [13,14]. The detailed MIPs synthesis process is described in Section 3.5.

Recently, molecularly imprinted polymers have aroused great interest and have been used in different fields, particularly for analytes extraction in complex samples. In the literature, just few published papers report the development of MIP-propyl paraben based and/or based on a limited selection of the above-cited molecules. Furthermore, such methods consider only environmental sample matrices. In the current study, following our research on innovative (micro)extraction procedures [15–25], different molecularly imprinted sorbent materials were studied for the extraction of most common parabens, including MPB, EPB, iPPB, PPB, iBPB, BPB, and BzPB. These sorbent phases were newly synthesized in our laboratory and herein tested in order to compare them in terms of major analytical parameters such as breakthrough volume (V_B), retention volume (V_R), hold-up volume (V_M), retention factor (k), and theoretical plates number (N). Furthermore, studies on selectivity have been carried out in this work in order to obtain a full characterization for these new sorbent materials when applied on real urine samples.

2. Results and Discussion

2.1. Experimental Determination of the Breakthrough Volume

The extraction capacity in the SPE depends on many factors such as the analyte retention capacity, the loaded sample volume, the conditioning solvents type, and volumes [26].

Nowadays, many new materials are available as sorbent material for SPE. To understand the extraction mechanism of these materials and to improve the extraction efficiency, it is important to characterize them. In SPE, the most important parameters that need to be calculated are breakthrough volume (V_B), retention volume (V_R), hold-up volume (V_M), retention factor (k), and theoretical plates number (N). For the determination of breakthrough volume, a solution containing the target analytes is continually applied to the SPE cartridge containing the sorbent material; breakthrough occurs when the capacity of sorbent has been exhausted [15,26,27].

The experimental curves (concentration vs. solution volume) were fitted by means of Boltzmann's function, and the regression parameters were used in calculating the parameters (breakthrough volume, retention volume, hold-up volume, retention factor, and theoretical plates number) for each considered stationary phases loaded on SPE cartridges [15,26,27]. Figure 2 reports the Boltzmann's functions by plotting the concentration of the analyte against the aliquot volumes passed through the sorbent for each type of MIPs and for all tested analytes. These graphs represent the dependence of parabens concentrations on successive 50 mL volume loading of aqueous samples. Table 2 reports the breakthrough volume (V_B), retention volume (V_R), hold-up volume (V_M), retention factor (k), and theoretical plates number (N) values calculated by using the proposed mathematical approach [15,26,27]. From the results presented in Table 2, it is evident that, as expected, the MIPs have a higher affinity for the molecule than the non-imprinted polymers (NIPs) and the polymer imprinted with BzPB has the highest breakthrough volumes, showing the highest capacity in trapping the analytes.

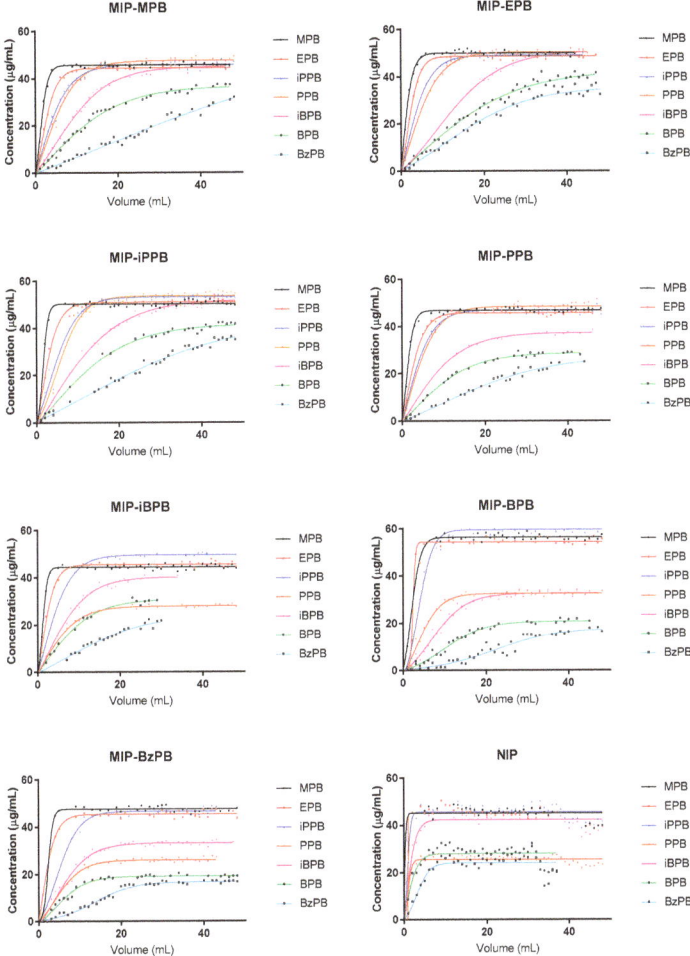

Figure 2. Boltzmann's functions determined for parabens on the different MIPs sorbent.

As reported by Bacalum et al. [27], the concentration of some compounds has different values at infinite time when passing through different MIP cartridges, and this finding could be explained by a change of interaction between analyte and the sorbent (e.g., two distinct adsorption profiles: the first herein observed, the second at higher sample volumes). Additionally, as reported by Bielica-Daszkiewicz [26] and Bacalum [27], it can be possible to observe curves where the maximum concentration used to test the sorbents not reached.

Furthermore, the curves obtained for the NIP show a shape change at very low volumes, highlighting that no retention mechanisms occur. Only for BzPB could be observed a little change, probably related to the different interactions, which could occur in relation to the chemical structure. In fact, in the presence of this analyte, larger breakthrough volume values are observed in all types of prepared MIPs, including NIPs. In this case, not only the interactions linked to the common basic structure for all parabens could be present but also there could be greater retention probably related to the second aromatic system (with related π–π interactions).

Table 2. Parameters determined for the analytes on different MIP sorbent at the concentration of 50 μg/mL.

Sorbent	Analyte	Breakthrough Volume (mL)	Hold-Up Volume (mL)	Retention Volume (mL)	Retention Factor (k)	Theoretical Plates (N)
MIP-MPB	Methyl paraben	2.19	6.78	1.1	2	5.18
	Ethyl paraben	2.1	11.81	1.06	2	10.15
	i-Propyl paraben	0.13	19.85	0.05	0	431.06
	Propyl paraben	2.78	23.93	1.37	2	16.51
	i-Butyl paraben	8.54	41.07	4.21	2	8.75
	Butyl paraben	0.91	50.47	0.53	5	94.1
	Benzyl paraben	3.41	273.36	1.57	1	173.12
MIP-EPB	Methyl paraben	0.53	6.69	0.27	2	24.17
	Ethyl paraben	2.51	10.13	1.26	2	7.07
	i-Propyl paraben	0.83	16.61	0.44	2.65	37.03
	Propyl paraben	1.84	22.22	0.92	2	23.23
	i-Butyl paraben	16.61	54.07	8.21	2	5.58
	Butyl paraben	3.54	69.81	1.59	1	42.99
	Benzyl paraben	20.46	70.82	10	2	6.08
MIP-iPPB	Methyl paraben	3.04	6	1.55	2	2.88
	Ethyl paraben	0.96	12.11	0.54	4	21.36
	i-Propyl paraben	6.13	22.64	2.97	1.64	6.63
	Propyl paraben	9.58	24.69	4.73	2	4.22
	i-Butyl paraben	1.96	47.96	1.15	5	40.6
	Butyl paraben	3.3	53.73	1.77	3	29.37
	Benzyl paraben	22.72	115.06	10.95	2	9.51
MIP-PPB	Methyl paraben	1.82	6.35	0.92	2	5.89
	Ethyl paraben	1.01	12.29	0.51	2	23.06
	i-Propyl paraben	0.56	17.24	0.29	2.59	58.4
	Propyl paraben	1.54	19.89	0.8	2	23.95
	i-Butyl paraben	0.84	34.91	0.46	3	74.51
	Butyl paraben	1.6	40.73	0.85	3	46.9
	Benzyl paraben	14.96	84.68	7.44	2	10.39

Table 2. Cont.

Sorbent	Analyte	Breakthrough Volume (mL)	Hold-Up Volume (mL)	Retention Volume (mL)	Retention Factor (k)	Theoretical Plates (N)
MIP-iBPB	Methyl paraben	2.42	5.13	1.21	2	3.25
	Ethyl paraben	1.06	9.85	0.56	3	16.74
	i-Propyl paraben	2.01	19.55	1	1.99	18.5
	Propyl paraben	0.41	23.83	0.22	3	105.23
	i-Butyl paraben	3.16	28.74	1.63	2	16.65
	Butyl paraben	2.14	33.26	1.06	2	30.38
	Benzyl paraben	13.18	62.6	6.63	2	8.44
MIP-BPB	Methyl paraben	3.8	9.39	1.9	2	3.96
	Ethyl paraben	4.18	5.59	2.35	4	1.37
	i-Propyl paraben	6.56	14.92	3.54	3.16	3.21
	Propyl paraben	5.56	19.85	2.99	3	5.63
	i-Butyl paraben	12.17	33.06	6.31	2	4.24
	Butyl paraben	19.14	43.04	9.52	2	3.52
	Benzyl paraben	54.34	88.58	22.22	1	2.99
MIP-BzPB	Methyl paraben	4.63	7.39	2.31	2	2.2
	Ethyl paraben	0.17	9.85	0.08	1	125.6
	i-Propyl paraben	8.58	21.77	4.28	1.96	4.09
	Propyl paraben	7.16	23.74	3.57	2	5.66
	i-Butyl paraben	6.62	27.27	3.55	3	6.67
	Butyl paraben	6.85	25.19	3.71	3	5.79
	Benzyl paraben	30.46	52.47	13.42	1	2.91
NIP	Methyl paraben	-	-	-	-	-
	Ethyl paraben	-	-	-	-	-
	i-Propyl paraben	1.71	3.61	0.86	1.98	3.22
	Propyl paraben	1.84	4.06	0.92	2	3.41
	i-Butyl paraben	0.11	6.69	0.06	3	109.55
	Butyl paraben	-	7.53	-	3	-
	Benzyl paraben	6.3	13.55	3.02	2	3.49

2.2. Selectivity Study of Molecularly Imprinted Polymers

Competitive adsorptions of methyl paraben, ethyl paraben, propyl paraben, *iso*-propyl paraben, propyl paraben, *iso*-butyl paraben, butyl paraben, and benzyl paraben were evaluated using imprinted and non-imprinted polymer. The Boltzmann's functions calculated for single paraben compared to all different molecular imprinted polymers herein tested, including NIP, were reported in Supplementary Materials, Section S.1, Figure S.1.1.

Furthermore, the selectivity that each paraben showed in its molecularly imprinted polymer compared with other imprinted polymers has been evaluated. The deviations could be related to the unselective interactions between the target analyte and the MIP phase. Table 3 shows data from the linear regression carried out to obtain comparative data of each single paraben in the various imprinted polymers, while in Supplementary Materials, the Figure S.1.2 reports the graphs for the single parabens. The values obtained for each paraben in the reference polymer have been placed on the *x*-axis, while on the *y*-axis the data relating to the single parabens in the polymers imprinted with different template are placed. This comparison was carried out in order to better highlight how the imprinted polymers are selective toward the single parabens with respect to other molecules with similar structure. By means of the obtained results, the imprinted polymers show a limited selectivity toward the single paraben, as can be observed from calculated slope values (nearest to 1 for all parabens). In particular, the imprinted polymers with PPB, iBPB, BPB, and BzPB template result more selective toward single parabens compared with MPB, EPB, and iPPB-imprinted polymers.

Table 3. Linear regression data.

Analyte	Sorbent	Slope	Y-Intercept	X-Intercept	1/Slope
MPB	MIP-MPB	1	0	0	1
	MIP-EPB	1.046 ± 0.025	1.85 ± 1.05	−1.77	0.96
	MIP-iPPB	1.134 ± 0.023	−1.29 ± 0.98	1.14	0.88
	MIP-PPB	0.999 ± 0.025	1.2 ± 1.1	−1.2	1.0
	MIP-iBPB	0.927 ± 0.033	2.6 ± 1.4	−2.8	1.1
	MIP-BPB	1.295 ± 0.037	−2.7 ± 1.6	2.07	0.77
	MIP-BzPB	1.039 ± 0.026	0.1.145	−0.12	0.96
	NIP	0.972 ± 0.064	0.2.82	−0.25	1.0
EPB	MIP-MPB	0.944 ± 0.027	−1.7 ± 1.1	1.8	1.1
	MIP-EPB	1	0	0	1
	MIP-iPPB	1.098 ± 0.027	−2.7 ± 1.2	2.50	0.91
	MIP-PPB	0.952 ± 0.026	−0.45 ± 1.2	0.48	1.05
	MIP-iBPB	0.952 ± 0.026	−0.45 ± 1.2	0.48	1.05
	MIP-BPB	1.214 ± 0.075	−3.3.305	2.75	0.82
	MIP-BzPB	0.898 ± 0.019	1.53 ± 0.84	−1.7	1.1
	NIP	0.604 ± 0.090	16.4 ± 4.1	−27.1	1.7
iPPB	MIP-MPB	0.803 ± 0.029	3.2 ± 1.4	−4.0	1.2
	MIP-EPB	0.907 ± 0.015	1.73 ± 0.69	−1.9	1.1
	MIP-iPPB	1	0	0	1
	MIP-PPB	0.853 ± 0.018	1.94 ± 0.89	−2.3	1.2
	MIP-iBPB	0.915 ± 0.014	1.00 ± 0.68	−1.1	1.1
	MIP-BPB	1.115 ± 0.050	1.2.412	−1.09	0.90
	MIP-BzPB	0.875 ± 0.029	0.06 ± 1.4	−0.07	1.1
	NIP	0.387 ± 0.085	25.6 ± 4.1	−66.2	2.6
PPB	MIP-MPB	1.011 ± 0.035	−2.2 ± 1.5	2.18	0.99
	MIP-EPB	1.045 ± 0.026	−0.5 ± 1.1	0.45	0.96
	MIP-iPPB	1.13 ± 0.022	−1.3 ± 1.0	1.14	0.88
	MIP-PPB	1	0	0	1
	MIP-iBPB	0.581 ± 0.017	−0.50 ± 0.78	0.85	1.7
	MIP-BPB	0.722 ± 0.021	−2.41 ± 0.96	3.3	1.4
	MIP-BzPB	0.541 ± 0.025	−0.25 ± 1.2	0.5	1.8
	NIP	0.282 ± 0.051	12.1 ± 2.3	−42.8	3.5

Table 3. Cont.

Analyte	Sorbent	Slope	Y-Intercept	X-Intercept	1/Slope
iBPB	MIP-MPB	1.022 ± 0.061	−1.7 ± 1.9	1.62	0.98
	MIP-EPB	1.021 ± 0.031	0.08 ± 1.00	−0.08	0.98
	MIP-iPPB	1.16 ± 0.11	−3.3.91	2.87	0.86
	MIP-PPB	0.884 ± 0.025	−0.25 ± 0.64	0.3	1.1
	MIP-iBPB	1	0	0	1
	MIP-BPB	0.827 ± 0.047	−2.2 ± 1.4	2.7	1.2
	MIP-BzPB	0.382 ± 0.052	−2.7 ± 1.7	7.2	2.6
	NIP	0.81 ± 0.12	15.1 ± 4.0	−18.7	1.2
BPB	MIP-MPB	1.521 ± 0.061	2.23 ± 0.81	−1.46	0.66
	MIP-EPB	1.723 ± 0.087	−0.3 ± 1.4	0.19	0.58
	MIP-iPPB	1.795 ± 0.060	1.3 ± 1.0	−0.72	0.56
	MIP-PPB	1.233 ± 0.056	2.97 ± 0.82	−2.41	0.81
	MIP-iBPB	1.46 ± 0.14	4.4 ± 1.4	−3.01	0.69
	MIP-BPB	1	0	0	1
	MIP-BzPB	0.919 ± 0.098	3.3 ± 1.1	−3.6	1.1
	NIP	0.55 ± 0.17	19.9 ± 2.4	−35.9	1.8
BzPB	MIP-MPB	1.40 ± 0.13	−1.1.645	0.76	0.72
	MIP-EPB	1.879 ± 0.061	0.69 ± 0.73	−0.37	0.53
	MIP-iPPB	1.78 ± 0.13	0.3 ± 1.8	−0.17	0.56
	MIP-PPB	1.220 ± 0.067	0.92 ± 0.80	−0.75	0.82
	MIP-iBPB	1.114 ± 0.064	2.66 ± 0.74	−2.39	0.90
	MIP-BPB	0.895 ± 0.081	−1.52 ± 0.98	1.7	1.1
	MIP-BzPB	1	0	0	1
	NIP	0.75 ± 0.18	13.8 ± 2.0	−18.5	1.3

However, some anomalies related to the PPB, i-BPB; BPB, and BzPB (chemical structures reported in Supplementary Materials, Figure S.1.3) have been found in MIP-PPB and MIP-BPB. MIP-PPB shows low retention capacity toward iBPB and it could be related to the chemical structure (-CH$_3$ instead -H). The –CH$_3$ group shows a little higher steric hindrance compared to only -H and this may partly limit the retention of this analyte for a still relatively small MIP cavity. This element can also be justified by the fact that, as shown in Table 3, the slope values are >1 for small parabens when the MIP was obtained with larger parabens as template. In addition, MIP-BPB had low retention capacity for BzPB; this is probably related to aromatic group in chemical structure of BzPB that increase the steric hindrance and not allow to a complete fit with the interaction folder.

For NIP, all values reported in Table 3 were below the unit, demonstrating that this non-imprinted polymer shows no selectivity against the considered analytes. As also highlighted in Table 2, NIP shows low breakthrough volume (mL) values in respect all the other MIPs. Furthermore, in Table 2 it was not possible to evaluate several parameters for the NIP as the Boltzmann's functions reported floating point errors.

From the analyses carried out, it appears evident that most of the selectivity of the phases is linked to the structure common to all the parabens (slope values reported in Table 3 next to unity) and that this is maximized when the same analyte is evaluated used as a different template (unitary slope). When the binding site is similar in size to the analyte, phase selectivity is not observed and interactions with the common paraben structure prevail. Similarly, if the site is larger, it is observed that structurally similar (but smaller in size) analytes are mostly retained (slope greater than 1).

For MPB the values are all close to the unit independently of the MIP. Similarly, similar behavior is observed for EPB and iPPB. Only from the PPB a distinction is made between selectivity based on the slope of the curves with values close to 1 for analytes of smaller dimensions and different from the unit (<1) for larger structures.

2.3. Real Sample Analysis

This procedure was applied to real urine sample from a healthy donor, who did not voluntarily take parabens or products containing parabens for analysis. The real sample was derived from our previous project [4], where only EPB was found in one urine real sample (as reported in Section 3.3). The urine sample, stored at −20 °C before analysis, was extracted following the MIP-SPE procedure reported in Section 3.7, and subsequently the eluate was injected in HPLC-PDA system. Using the MIP-EPB as sorbent in the SPE procedure, an improvement of S/N ratio of 3–5 times has been observed compared with previously obtained ratio [4], confirming the sensibility of imprinting polymers for this class of compounds. To fully compare fabric phase sorptive extraction-HPLC-PDA (FPSE-HPLC-PDA) vs. MIP-SPE-HPLC-PDA, a partial validation for MIP-SPE based procedure would be necessary, and then to compare the results by mean of statistical approach. From the point of view of the comparison between these 2 configurations, it would, however, be correct to provide for the validation of the new MIP-SPE-HPLC-PDA configuration (in order to have the best and optimized performances) and subsequently it can be applied a t-test.

2.4. MIPs Characterization

The evaluation of the chemical–physical characteristics of the created MIPs is reported in the Supplementary Materials, Section S.1.4. From the data obtained through Fourier transform infrared spectroscopy (FTIR), differential thermal analysis (DTA), and thermogravimetric (TGA) analysis it is possible to highlight the reproducibility of the synthesis process using different templates.

3. Materials and Methods

3.1. Chemicals, Solvents, and Devices

The International Forensic Research Institute, Department of Chemistry and Biochemistry, Florida International University (Miami, FL, USA) provided all parabens chemical standards (methyl paraben, ethyl paraben, propyl paraben, isopropyl paraben, butyl paraben, isobutyl paraben, and benzyl paraben) and all molecularly imprinted polymers.

Sodium phosphate monobasic, sodium phosphate dibasic (>99% purity grade), and phosphoric acid were obtained from Sigma-Aldrich (Milan, Italy). Acetonitrile and methanol (HPLC-grade) were purchased from Honeywell (New Jersey, USA) and were used without further purification. Deionized water (18.2 MΩ-cm at 25 °C) was generated by a Millipore MilliQ Plus water (Millipore Bedford Corp., Bedford, MA, USA). GraphPad Prism v.4 was used for the statistical analysis of experimental data.

3.2. Stock Solution and Working Solution Preparation

Stock solutions of chemical standards were prepared in methanol at the concentration of 1 mg/mL. The working solutions were prepared by dilution of stock solution in Milli-Q water and stored at 4 °C. The resulting samples were used to evaluate the enrichment factors, breakthrough volume, retention volume, hold-up volume, retention factor, and theoretical plates number.

3.3. Human Urine Sample Collection, Storage, and Preparation

Human urine sample was collected from a healthy volunteer informed about the nature of the study, who did not voluntarily take parabens or products containing parabens for analysis purposes. This urine sample is the same (a second aliquot) analyzed in the previous study [4], stored at −20 °C before analysis.

3.4. Apparatus and Chromatographic Conditions

The method has been described and validated by Tartaglia et al. [4]. Briefly, analyses were performed using an HPLC Thermo Fisher Scientific liquid chromatography system (Model: Spectra

System P2000) coupled to a photodiode array detector (PDA) Model: Spectra System UV6000LP. Mobile phase was directly on-line degassed by using a Spectra System SCM1000 (Thermo Fisher Scientific, Waltham, MA, USA). Excalibur v.2.0 software (Thermo Fisher Scientific, Waltham, MA, USA) was used to collect and analyze data. Spherisorb C18 (15 cm × 4.6 mm, 5 µm) was used to resolve all parabens; the column was thermostated at 27 °C (± 1 °C) using a Jetstream2 Plus column oven during the analysis. The chromatographic separation was conducted in isocratic elution using phosphate buffer (28 mM, pH = 2.5) as solvent A and methanol as solvent B in volume percentages of 55 and 45, respectively. The flow rate was set at 1 mL/min. All the compounds were detected at the maximum wavelengths of 257 nm with retention time of 3.97, 6.00, 8.83, 10.43, 18.37, 19.75, and 22.33 min (for MPB, EPB, iPPB, PPB, iBPB, BPB, and BzPB, respectively).

3.5. Synthesis of Novel MIPs-Parabens and NIP

Preparation of parabens imprinted polymers involve several distinct steps: *(i)* complexation of parabens with sol–gel functional precursors; *(ii)* hydrolysis of the sol–gel cross-linking reagent; *(iii)* condensation of hydrolyzed sol–gel cross-linking reagent in presence of paraben-functional precursor complex; *(iv)* removal of paraben templates from paraben imprinted sorbents; and *(v)* synthesis of non-imprinted polymer (NIP) sorbents.

3.5.1. Complexation of Parabens with Sol–Gel Functional Precursors

The complexation of individual paraben with sol–gel precursors is a spontaneous, self-assembling process directed by intermolecular interactions, e.g., hydrogen bonding, π–π interactions, and van der Waals force between the paraben template and sol–gel functional precursors. To achieve superior specificity, two sol–gel precursors phenyltriethoxysilane (PTES) and 3-aminopropyl triethoxysilane (3-APTES) were employed in the complexation process. The complexation was carried out by sequentially adding the paraben template:ethanol:PTES:3-APTES in a 50-mL centrifuge tube at a molar ratio 1:80:3:3, respectively. The mixture was vortexed vigorously after adding each ingredient. Subsequently, the solution was sonicated for 30 min. The mixture was then incubated at room temperature for 6 h so that the sol–gel functional precursors self-assembled themselves around the template molecules by hydrogen bonding and π–π interactions.

3.5.2. Hydrolysis of Sol–Gel Cross-Linking Reagent

Tetramethyl orthosilicate (TMOS) was used as the sol–gel cross-linking reagent. Four methoxy functional groups connected to the central silica atom must be hydrolyzed first so that they can undergo polycondensation to build the inorganic silica network. Hydrolysis of TMOS was carried under acidic condition using trifluoroacetic acid (TFA) as the acid catalyst. The molar ratio of the template:TMOS:ethanol:TFA:water was maintained at 1:15:45:0.1:55, respectively. The mixture was prepared in a 50-mL centrifuge tube by adding individual ingredient, vortexing for 5 min and finally sonicating for 30 min. The sol solution was kept in a silicon bath at 50 °C for 12 h to ensure complete hydrolysis of TMOS in presence of the acid catalyst.

3.5.3. Condensation of Hydrolyzed Sol–Gel Cross-Linking Reagent in Presence of the Paraben-Functional Precursors Complex

To create paraben imprinted polymers, paraben sol–gel functional precursors complexes are added to hydrolyzed sol solution in droplets under continuous stirring on a magnetic stirrer. During this process, paraben complexes randomly orient themselves within the growing sol–gel network with minimal steric hindrance and subsequently become frozen with the 3D sol–gel network. In order to complete and expedite the network formation process via condensation, the sol–gel polymer was kept in a silicon oil bath at 50 °C for 24 h. During this residence time in the oil bath, aging and ripening of the sol–gel network occurs, leading to a robust, highly porous silica network with trapped solvent, templates, and unreacted sol solution ingredients.

3.5.4. Removal of Paraben Templates from the Imprinted Polymers

Quantitative and exhaustive removal of paraben templates from the imprinted polymer is one of the most important and challenging tasks in the creation of parabens imprinted polymers. Successful removal of the templates leave a nanocavity complimentary to the size, shape, and functionality of the template molecules and creates a highly specific synthetic receptor site, extremely affinitive towards the template molecules. To remove the templates, the paraben imprinted sol–gel polymers were first dried at 100 °C in a vacuum drier for 48 h. The dried MIPs are then crushed and ground in a mortar into fine particles (~50 μm) and subsequently subjected to accelerated solvent extraction at 150 °C and 1500 psi for 30 min using methanol as the extraction solvent. Additionally, during the analyses for breakthrough volume curves, no parabens release was observed in the conditioning steps, highlighting that the phases were completely cleaned during accelerated solvent extraction (ASE) process.

3.5.5. Synthesis of Non-Imprinted Polymers

The success of molecular imprinting is often evaluated by comparing the adsorption of the template molecules on the imprinted polymer bed over a non-imprinted polymer (NIP) bed under identical conditions. Nonspecific adsorption of both the MIP and NIP can be easily estimated by exposing the known mass of the polymers into a known volume of an aqueous solution containing a known concentration of the template molecules. As such, paraben non-imprinted polymers were synthesized in parallel to the MIPs using identical process and sol solution with the only exception that no paraben templates were added in the NIP materials.

3.6. Preparation of MIP–SPE Media

A frit was placed on the bottom of an empty SPE cartridge (1 mL in total volume), which was employed as the MIP–SPE column. The SPE cartridges used in this study were in-house packed with 30 mg of parabens-MIPs (the stationary phase synthesized as previously reported in Section 3.5). Then, another frit was placed on the top of the cartridge.

3.7. Molecularly Imprinted Polymer–Solid Phase Extraction Procedure

For the breakthrough curve construction, SPE experiment was carried out as follows: 50 mL of solution at 50 μg/mL was introduced into SPE sorbent (1 mL per time), conditioned with methanol (1 mL) and water (1 mL). Each sample aliquot (1 mL) was collected in a separate vial and analyzed by HPLC-PDA to measure the concentration of analyte in each aliquot of sample. Breakthrough curves were determined as the relationship between concentration after extraction and the total volume passed through the sorbent.

For the analysis of the urine sample, the SPE cartridge loaded with molecularly imprinted polymer was conditioned by 1 mL of methanol and followed 1 mL of Milli-Q water. 1 mL of sample solution was passed through the cartridge at a flow of 1 mL/min. Subsequently, the cartridge was consecutively rinsed with 2 mL of water, and 3 mL of MeOH was used as elution solvent. Elutes (20 μL) are then injected in HPLC-PDA system.

4. Conclusions

A novel, MIP sorbent based on seven parabens template were synthesized. The new MIP-SPE method coupled with HPLC-PDA analysis was preliminarily tested for the determination of parabens in human urine, confirming that molecular imprinting technology represents a valid strategy for the synthesis of new selective extraction materials. All molecularly imprinted polymers have been characterized in terms of extraction capability, breakthrough volume, retention volume, hold-up volume, number of theoretical plates, and retention factor.

Furthermore, selectivity studies have been performed, comparing the extraction efficiency of each molecularly imprinted polymer for all the parabens tested in this study. Comparing the calculated parameters, the imprinted polymer does not show marked selectivity for every single paraben, but they provided "*class-specific*" interactions. Following this work, a polymer based on a mixture of individual MIPs for the extraction of these compounds from complex matrices could be developed. Greater efforts are therefore needed in future studies to obtain extractive materials based on imprinted polymers with increased selectivity and specificity.

Supplementary Materials: The following are available online at http://www.mdpi.com/1420-3049/24/18/3334/s1, Figure S.1.1. Boltzmann's functions for single parabens., Figure S.1.2. Selectivity of each paraben., Figure S.1.3. Chemical structures of **a.** Propyl paraben; **b.** *iso*–Butyl paraben; **c.** Butyl paraben; **d.** Benzyl paraben. All the axes report the concentration values (µg/mL).

Author Contributions: Methodology, M.L., A.K., K.G.F., H.I.U., F.C., and G.M.M.; formal analysis, S.U., A.T., P.B., F.M., and S.G.; investigation, A.T., P.B., and F.M.; data curation, A.T.; writing—original draft preparation, A.T. and M.L.; writing—review and editing, M.L. and A.K.; supervision, M.L., A.K., K.G.F., H.I.U., C.D.O., U.d.G., and F.C, D.M.; project administration, M.L., A.K., K.G.F., F.S., H.I.U., C.D.O., U.d.G., F.C., and D.M.

Funding: This research received no external funding.

Acknowledgments: This work was supported by grant MIUR ex 60%, University of Chieti – Pescara "G. d'Annunzio", Chieti, Italy. Authors want to thank TUBITAK 2219 RESEARCH SCHOLARSHIP PROGRAM for supporting the researches of Songul Ulusoy. Authors want to thank Enrico Marcantoni at the Chemical Science Department—School of Science and Technology at the University of Camerino (Italy) for making available the FTIR and TGA instrumentations. Authors want to thank Mauro Medoro of University "G. d'Annunzio", Chieti (Italy) for his kind help in all the experiments.

Conflicts of Interest: The authors declare no conflict of interest.

References

1. Piao, C.; Chen, L.; Wang, A. A review of the extraction and chromatographic determination methods for the analysis of parabens. *J. Chromatogr. B* **2014**, *969*, 139–148. [CrossRef] [PubMed]
2. Ocaña-González, J.A.; Villar-Navarro, M.; Ramos-Payán, M.; Fernández-Torres, R.; Bello-López, M.A. New developments in the extraction and determination of parabens in cosmetics and environmental samples. A review. *Anal. Chim. Acta* **2015**, *858*, 1–15. [CrossRef] [PubMed]
3. Giulivo, M.; De Alda, M.L.; Capri, E.; Barceló, D. Human exposure to endocrine e disrupting compounds: Their role in reproductive systems, metabolic syndrome and breast cancer. A review. *Environ. Res.* **2016**, *151*, 251–264. [CrossRef] [PubMed]
4. Tartaglia, A.; Kabir, A.; Ulusoy, S.; Sperandio, E.; Piccolantonio, S.; Ulusoy, H.I.; Furton, K.G.; Locatelli, M. FPSE-HPLC-PDA analysis of seven paraben residues in human whole blood, plasma, and urine. *J. Chromatogr. B* **2019**, 1125. [CrossRef] [PubMed]
5. Ye, X.; Kuklenyik, Z.; Bishop, A.M.; Needham, L.L.; Calafat, A.M. Quantification of the urinary concentrations of parabens in humans by on-line solid phase extraction-high performance liquid chromatography–isotope dilution tandem mass spectrometry. *J. Chromatogr. B* **2006**, *844*, 53–59. [CrossRef] [PubMed]
6. Gonzalez-Marino, I.; Quintana, J.B.; Rodrıguez, I.; Cela, R. Simultaneous determination of parabens, triclosan and triclocarban in water by liquid chromatography/electrospray ionization tandem mass spectrometry. *Mass Spectrom.* **2009**, *23*, 1756–1766. [CrossRef]
7. Blanco, E.; Casais, M.d.C.; Mejuto, M.d.C.; Cela, R. Combination of off-line solid-phase extraction and on-column sample stacking for sensitive determination of parabens and p-hydroxybenzoic acid in waters by non-aqueous capillary electrophoresis. *Anal. Chim. Acta* **2009**, *647*, 104–111. [CrossRef]
8. Zotou, A.; Sakla, I.; Tzanavaras, P.D. LC-determination of five paraben preservatives in saliva and toothpaste samples using UV detection and a short monolithic column. *J. Pharm. Biomed. Anal.* **2007**, *53*, 785–789. [CrossRef]
9. Regueiro, J.; Becerril, E.; Garcia-Jares, C.; Llompart, M. Trace analysis of parabens, triclosan and related chlorophenols in water by headspace solid-phase microextraction with in situ derivatization and gas chromatography–tandem mass spectrometry. *J. Chromatogr. A* **2009**, *1216*, 4693–4702. [CrossRef]
10. Sanagi, M.M.; Salleh, S.; Ibrahim, W.A.W.; Naim, A.A.; Hermawan, D.; Miskam, M.; Hussain, I.; Aboul-Enein, H.Y. Molecularly imprinted polymer solid-phase extraction for the analysis of organophosphorus pesticides in fruit samples. *J. Food Compos. Anal.* **2012**, *32*, 155–161. [CrossRef]

11. Li, G.; Row, K.H. Recent applications of molecularly imprinted polymers (MIPs) on micro-extraction techniques. *Sep. Purif. Rev.* **2018**, *47*, 1–18. [CrossRef]
12. Chen, L.; Xu, S.; Li, J. Recent advances in molecular imprinting technology: Current status, challenges and highlighted applications. *Chem. Soc. Rev.* **2011**, *40*, 2922–2942. [CrossRef] [PubMed]
13. Lofgreen, J.E.; Ozin, G.A. Controlling morphology and porosity to improve performance of molecularly imprinted sol-gel silica. *Chem. Soc. Rev.* **2014**, *43*, 911–933. [CrossRef] [PubMed]
14. Diaz-Garcia, M.E.; Laino, R.B. Molecular Imprinting in sol-gel materials: Recent developments and applications. *Microchim. Acta* **2005**, *149*, 19–36. [CrossRef]
15. Tartaglia, A.; Locatelli, M.; Kabir, A.; Furton, K.G.; Macerola, D.; Sperandio, E.; Piccolantonio, S.; Ulusoy, H.I.; Maroni, F.; Bruni, P.; et al. Comparison between Exhaustive and Equilibrium Extraction using different SPE Sorbents and Sol-gel Carbowax 20M Coated FPSE Media. *Molecules* **2019**, *24*, 382. [CrossRef] [PubMed]
16. Malatesta, L.; Cosco, D.; Paolino, D.; Cilurzo, F.; Costa, N.; Di Tullio, A.; Fresta, M.; Celia, C.; Di Marzio, L.; Locatelli, M. Simultaneous quantification of Gemcitabine and Irinotecan hydrochloride in rat plasma by using high performance liquid chromatography-diode array detector. *J. Pharm. Biomed. Anal.* **2018**, *159*, 192–199. [CrossRef]
17. Campestre, C.; Locatelli, M.; Guglielmi, P.; De Luca, E.; Bellagamba, G.; Menta, S.; Zengin, G.; Celia, C.; Di Marzio, L.; Carradori, S. Analysis of imidazoles and triazoles in biological samples after MicroExtraction by Packed Sorbent. *J. Enzyme Inhibit. Med. Chem.* **2017**, *32*, 1053–1063. [CrossRef]
18. Locatelli, M.; Kabir, A.; Innosa, D.; Lopatriello, T.; Furton, K.G. A Fabric Phase Sorptive Extraction-High Performance Liquid Chromatography-Photo Diode Array Detection Method for the Determination of Twelve Azole Antimicrobial Drug Residues in Human Plasma and Urine. *J. Chromatogr. B* **2017**, *1040*, 192–198. [CrossRef]
19. D'Angelo, V.; Tessari, F.; Bellagamba, G.; De Luca, E.; Cifelli, R.; Celia, C.; Primavera, R.; Di Francesco, M.; Paolino, D.; Di Marzio, L.; et al. MicroExtraction by Packed Sorbent and HPLC-PDA quantification of multiple anti-inflammatory drugs and fluoroquinolones in human plasma and urine. *J. Enzyme Inhibit. Med. Chem.* **2016**, *31*, 110–116. [CrossRef]
20. Locatelli, M.; Ciavarella, M.T.; Paolino, D.; Celia, C.; Fiscarelli, E.; Ricciotti, G.; Pompilio, A.; Di Bonaventura, G.; Grande, R.; Zengin, G.; et al. Determination of Ciprofloxacin and Levofloxacin in Human Sputum Collected from Cystic Fibrosis Patients using Microextraction by Packed Sorbent-High Performance Liquid Chromatography Photo Diode Array Detector. *J. Chromatogr. A* **2015**, *1419*, 58–66. [CrossRef]
21. Locatelli, M.; Cifelli, R.; Di Legge, C.; Barbacane, R.C.; Costa, N.; Fresta, M.; Celia, C.; Capolupo, C.; Di Marzio, L. Simultaneous determination of Eperisone Hydrochloride and Paracetamol in mouse plasma by High Performance Liquid Chromatography-PhotoDiode Array Detector. *J. Chromatogr. A* **2015**, *1388*, 79–86. [CrossRef] [PubMed]
22. Locatelli, M.; Furton, K.G.; Tartaglia, A.; Sperandio, E.; Ulusoy, H.I.; Kabir, A. An FPSE-HPLC-PDA method for rapid determination of solar UV filters in human whole blood, plasma and urine. *J. Chromatogr. B* **2019**, *1118*, 40–50. [CrossRef] [PubMed]
23. Locatelli, M.; Ferrone, V.; Cifelli, R.; Barbacane, R.C.; Carlucci, G. Microextraction by packed sorbent and high-performance liquid chromatography determination of seven non-steroidal anti-inflammatory drugs in human plasma and urine. *J. Chromatogr. A* **2014**, *1367*, 1–8. [CrossRef] [PubMed]
24. Locatelli, M.; Tinari, N.; Grassadonia, A.; Tartaglia, A.; Macerola, D.; Piccolantonio, S.; Sperandio, E.; D'Ovidio, C.; Carradori, S.; Ulusoy, H.I.; et al. FPSE-HPLC-DAD method for the quantification of anticancer drugs in human whole blood, plasma, and urine. *J. Chromatogr. B* **2018**, *1095*, 204–213. [CrossRef] [PubMed]
25. Kabir, A.; Furton, K.G.; Tinari, N.; Grossi, L.; Innosa, D.; Macerola, D.; Tartaglia, A.; Di Donato, V.; D'Ovidio, C.; Locatelli, M. Fabric phase sorptive extraction-high performance liquid chromatography-photo diode array detection method for simultaneous monitoring of three inflammatory bowel disease treatment drugs in whole blood, plasma and urine. *J. Chromatogr. B* **2018**, *1084*, 53–63. [CrossRef] [PubMed]
26. Bielica-Daszkiewicz, K.; Voelkel, A. Theoretical and experimental methods of determination of the breakthrough volume of SPE sorbents. *Talanta* **2009**, *80*, 614–621. [CrossRef]

27. Bacalum, E.; Tanase, A.; David, V. Retention mechanisms applied in solid phase extraction for some polar compounds. *An. Univ. Bucar. Chim.* **2010**, *19*, 61–68.

Sample Availability: Samples of the compounds are available from the authors.

© 2019 by the authors. Licensee MDPI, Basel, Switzerland. This article is an open access article distributed under the terms and conditions of the Creative Commons Attribution (CC BY) license (http://creativecommons.org/licenses/by/4.0/).

Article

Screen-printed Microsensors Using Polyoctyl-thiophene (POT) Conducting Polymer As Solid Transducer for Ultratrace Determination of Azides

Ahmed Galal Eldin [1], Abd El-Galil E. Amr [2,3,*], Ayman H. Kamel [1,*] and Saad S. M. Hassan [1,*]

1. Chemistry Department, Faculty of Science, Ain Shams University, 11566 Abbasia, Cairo, Egypt; ahmeddna2006@yahoo.com
2. Pharmaceutical Chemistry Department, Drug Exploration & Development Chair (DEDC), College of Pharmacy, King Saud University, Riyadh 11451, Saudi Arabia
3. Applied Organic Chemistry Department, National Research Centre, 12622 Dokki, Giza, Egypt
* Correspondence: aamr@ksu.edu.sa (A.E.-G.E.A.); ahkamel76@sci.asu.edu.eg (A.H.K.); saadsmhassan@yahoo.com (S.S.M.H.); Tel.: +966-565-148-750 (A.E.-G.E.A.); +20-1000-743-328 (A.H.K.); +20-1222-162-766 (S.S.M.H.)

Academic Editors: Clinio Locatelli and Marcello Locatelli
Received: 8 March 2019; Accepted: 2 April 2019; Published: 9 April 2019

Abstract: Two novel all-solid-state potentiometric sensors for the determination of azide ion are prepared and described here for the first time. The sensors are based on the use of iron II-phthalocyanine (Fe-PC) neutral carrier complex and nitron-azide ion-pair complex (Nit-N_3^-) as active recognition selective receptors, tetradodecylammonium tetrakis(4-chlorophenyl) borate (ETH 500) as lipophilic cationic additives and poly(octylthiophene) (POT) as the solid contact material on carbon screen-printed devices made from a ceramic substrate. The solid-contact material (POT) is placed on a carbon substrate (2 mm diameter) by drop-casting, followed, after drying, by coating with a plasticized PVC membrane containing the recognition sensing complexes. Over the pH range 6-9, the sensors display fast (< 10 s), linear potentiometric response for 1.0×10^{-2}–1.0×10^{-7} M azide with low detection limit of 1.0×10^{-7} and 7.7×10^{-8} M (i.e., 6.2–4.8 ng/ml) for Fe-PC/POT/and Nit-N_3^-/POT based sensors, respectively. The high potential stability and sensitivity of the proposed sensors are confirmed by electrochemical impedance spectroscopy (EIS) and constant-current chronopotentiometry measurement techniques. Strong membrane adhesion and absence of delamination of the membrane, due to possible formation of a water film between the recognition membranes and the electron conductor are also verified. The proposed sensors are successfully applied for azide quantification in synthetic primer mixture samples. Advantages offered by these sensors are the robustness, ease of fabrication, simple operation, stable potential response, high selectivity, good sensitivity and low cost.

Keywords: azides; screen-printed sensors; poly (3-octylthiophene); solid contact potentiometric sensors; iron-phthalocyanine; nitron-azide complexes

1. Introduction

Coated wire potentiometric electrodes (CWEs) are prepared by direct coating of metallic conductors such as platinum, copper, silver, gold and carbon substrates with electroactive species incorporated in a thin polymeric film. These devices are simple, inexpensive, do not need internal reference solutions, are easily miniaturized, and find application in chemical, biomedical and clinical analyses. The development of chemical field-effect transistors (CHEMFETs) was considered a logical extension of these devices [1,2]. However, these sensors gave low potential stability and poor reproducibility due to the direct contact betwwen two phases of quite different properties. The observed instability is attributed to the formation of a thin electrolyte solution layer between the conductive support and the membrane with composition changing in course of the measurement [3].

Recently, solid-contact potentiometric ion-selective sensors (SC-ISEs) based on the use of conducting polymers (CPs) have received considerable attention [4,5] and demonstrated a wide range of applications. Among the many conducting polymers that have been investigated, polypyrrole, poly(3-octylthiophene) (POT), polyaniline, and poly(3,4-ethylenedioxythiophene) (PEDOT) are the most commonly used polymers [4,5]. POT is less subject to reactions with ambient species such as oxygen because it has a relatively high oxidation potential and is usually used in an undoped ion-free form. Thus POT films have a relatively low redox capacitance and electronic conductivity. Ion-to-electron transduction at the interfaces occurs mainly through the electrical double layer that forms at the solid contact/membrane boundary [5].

This configuration seems ideal for fabrication of miniaturized sensors with sufficient analytical performance for trace analysis of some metal ions [6–9]. Unlike the conventional designs of potentiometric sensors, solid contact sensors are compatible with mass production technologies, which drive down unit cost and improving batch- to - batch reproducibility. On the other hand, the screen-printing technique has been successfully applied to mass production of low-cost, highly reproducible and reliable disposable sensors for rapid assessment of many types of analytes [10–12].

The present study deals with fabrication and application of solid contact potentiometric sensors with conducting polymer for sensitive and selective determination of azides which are widely used in explosive detonators, electrical discharge tubes, anti-corrosion solutions, production of foam rubber, laboratories preservatives, agricultural pest control, and automobile air bags [13]. On the other hand, azides are considered as potent toxins, similar to cyanides as both cause death due strongly binding to iron in hemoglobin [14]. An intake of about 10 µg/g azide ions causes death within half an hour [15]. In addition, it is readily protonated in the aquatic environment to yield volatile hydrazoic acid (HN_3) that can create an airborne hazard [16]. Therefore, tight controls on the allowed levels of azides in wastewater effluents, industrial solid waste and propellants are highly demanded.

Few azide-potentiometric polymeric membrane and gas sensors have been described [17–26]. Some of these devices exhibit narrow working concentration ranges and suffer from interference from various anions such as ClO_4^-, SO_4^{2-}, HCO_3^-, Cl^-, HPO_4^{2-}, and NO_3^-. In order to cope with these limitations, further efforts are required to develop new designs of potentiometric probes with lower detection limits, good selectivity, and high potential stability for azide monitoring.

In this work, novel solid contact potentiometric sensors using carbon screen-printed substrates were prepared, optimized and examined. These sensors are based on the use of $Fe^{II}PC$ ionophore and Nit-N_3^- ion association complex as sensing materials for azide ions. These sensors are applicable for trace analysis of azide ion. The performance characteristics of these sensors were evaluated and satisfactorily used for accurate determination of microgram quantities of azide. The sensors offered excellent advantages such as miniaturization, cost-effectiveness, ease of fabrication and high potential stability and sensitivity.

2. Results

2.1. Sensors Construction and Characteristics

Two all-solid state potentiometric microsensors were prepared, characterized and evaluated. These consist of a carbon screen printed planar ceramic substrate (2 × 2 mm) coated with poly- (octylthiophene) (POT) as a solid conducting layer and covered with a film of either iron(II)-phthalocyanine complex (FePC) or nitron-azide ion-pair complex (Nit-N_3^-) dispersed in the plasticized PVC film as active recognition selective receptors (Figure 1). The sensor cocktail consists of PVC, sensing ionophore, ETH 500 and the plasticizer in an optimum ratio of 32.2, 2.5,

2.0 and 63.3 wt%, respectively. ETH 500 was used as ion excluder and o-NPOE as a membrane plasticizer [27,28]. The plasticizer was selected to possess high lipophilicity, high molecular weight, low tendency for exudation from the membrane matrix, low vapor pressure and high capacity to dissolve the membrane ingradients. The use of ETH 500 additive in the membrane provided significant effect on the sensor response [29]. o-NPOE (ε_r = 24, M.wt. 435), DBS (ε_r = 5.1, M.wt. 390), and DOP (ε_r = 8.4, M.wt. 390.5 plasticizers were tested.

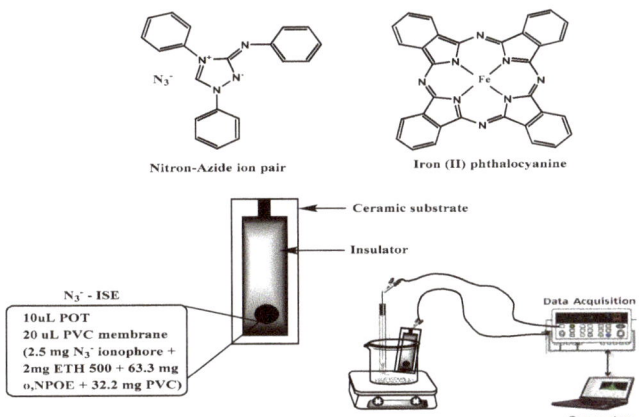

Figure 1. Chemical structures of azide ionophores used for azide membrane sensors and a schematic presentation of the proposed device.

The calibration plots of membrane sensors containing these solvent mediators are shown in Figure 2. Linear responses for the concentration ranges of 3.5×10^{-7}–1.0×10^{-2}, 1.0×10^{-6}–1.0×10^{-2} and 2.6×10^{-6}–1.0×10^{-2} M, and detection limits of 1.0×10^{-7}, 4.3×10^{-7} and 7.2×10^{-7} M with calibration slopes of -58.3 ± 0.9, -41.3 ± 0.6 and -41.4 ± 0.2 mV/decade for were obtained with FePC/POT based membranes plasticized with o-NPOE, DOP and DBS, respectively. Nit-N_3/POT based membrane sensors plasticized with o-NPOE, DOP and DBS, displayed calibration slopes of -55.1 ± 0.7, 48.2 ± 0.6 and -43.6 ± 0.5 mV/decade, linear responses over the concentration ranges of 1.0×10^{-7}–1.0×10^{-2}, 1.0×10^{-6}–1.0×10^{-2} and 1.0×10^{-6}–1.0×10^{-2} M, and detection limits 7.7×10^{-8}, 2.1×10^{-7} and 2.4×10^{-7} M, respectively.

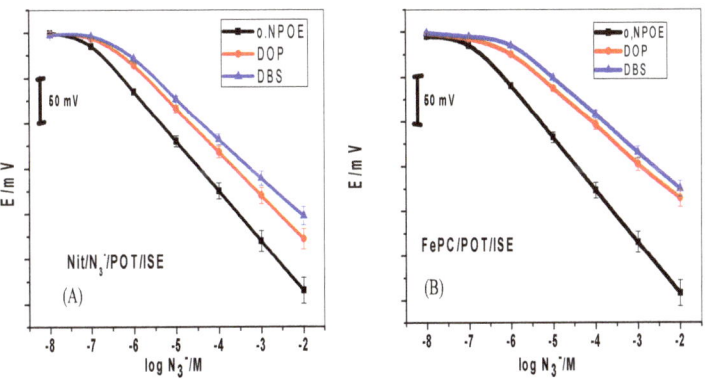

Figure 2. Effect of plasticizer polarity on the potentiometric plot of (**A**) Nit/N_3^-/POT-ISE and (**B**) FePC/POT-ISE. Background: 10^{-2} M Tris buffer solution, pH 7.0.

The performance characteristics for all these sensors are shown in Table 1. These results revealed that the sensors based on the use of o-NPOE as a solvent mediator displayed much better performance towards azide determination than other sensors based on DBS and DOP. Table 2 shows a comparison of the general performance of the proposed solid-contact azide sensors with those previously reported.

2.2. Robustness

The sensitivity of the proposed method to variations of experimental conditions (temperature, pH, and sample size) was tested. The ruggedness test was done using "Youden and Steiner partial factorial design" where eight replicate analyses were conducted, and three factors are varied and analyzed.

The pH effect on the potentiometric response of the proposed sensors was tested over a pH range of 2 to 11 with two fixed sodium azide concentrations (1.0×10^{-4} and 1.0×10^{-3} M). The potential response was pH independent in the pH range 5–10 and 6–9 for Nit-N_3/POT and FePC/POT based membrane sensors, respectively. At pH > 10, the potential decreased probably due to competition of the OH^- with the N_3^- anion. At pH < 5, a positive potential drift was noticed indication a decrease in azide concentration due to the formation and volatilization of hydrazoic acid (HN_3) gas.

Variation of the concentration of azide samples over the range 10^{-5}–10^{-2} M did not affect the accuracy by more than 1%. Change of the temperature of the test solution from 18–25 °C slightly affected the results. The simplest form of the Nernst equation is: $E = E° - (0.055/n) \log c$. However, the 0.055/n part of the equation is a simplification of $2.303RT/nF$. So, at 18 °C, $2.303RT/F = 0.053$ volts and upon increasing the temperature to 25 °C, this value goes up to 0.055 volts.

The response time and stability of the proposed sensors were measured by recording the time required by the two sensors to reach a stable steady-state potential (within ± 0.3 mV). A response time of less than 10 s was obtained for all azide solutions in the linear calibration range of the two examined sensors. Potential stability was tested by following within-day repeatability and between-days reproducibility of the potentiometric response of the sensors. The potentials response remained constant within ± 0.3 mV for at least 30 min. The results obtained with 6 identical sensors, prepared and used over a period of 2 months showed a standard deviation not exceeding ± 1.1 mV without observing any considerable changes in the selectivity and response time. The reproducibility of the calibration slope was within ± 1.5 mV/decade over a period of 8 weeks ($n = 6$). After two months, the detection limit gradually changed about half an order of magnitude.

Table 1. Response characteristics of azide membrane sensors in 0.01 M Tris buffer of pH 7.

Parameter	FePC/POT			Nit-N₃/POT		
	o-NPOE	DOP	DBS	o-NPOE	DOP	DBS
Slope, (mV/decade)	−58.3 ± 0.9	−41.3 ± 0.6	−41.4 ± 0.2	−55.1 ± 0.7	−48.2 ± 0.6	−43.6 ± 0.5
Coefficient, (r) (n=3)	−0.998	−0.997	−0.999	−0.998	−0.997	0.999
Detection limit, (M)	1.0×10^{-7}	4.3×10^{-7}	7.2×10^{-7}	7.7×10^{-8}	2.1×10^{-7}	2.4×10^{-7}
Linear range, (M)	3.5×10^{-7}–1.0×10^{-2}	1.0×10^{-6}–1.0×10^{-2}	2.6×10^{-6}–1.0×10^{-2}	1.0×10^{-7}–1.0×10^{-2}	1.0×10^{-6}–1.0×10^{-2}	1.0×10^{-6}–1.0×10^{-2}
Response time, (s)	<10	<10	<10	<10	<10	<10
Working range, (pH)	5.0–10	5.0–10	5–10	6.0–9	6.0–9	6.0–9
Standard deviation, (%)	0.7	1.3	1.1	0.8	0.5	0.7
Accuracy, (%)	99.6	99.3	98.8	98.4	99.5	99.3
Precision, (%), Cv_W (%)	1.1	1.2	1.7	0.7	1.0	1.2

Table 2. General features of some potentiometric azide membrane sensors based on different ionophores.

Ionophore	Slope, (mV/decade)	Linear Range, (M)	pH Range	Detection Limit, (M)	Interference	Ref.
Fe^III- and Co^III-complexes of 2,3,7,8,12,13,17,18-octakis (benzylthio)-5,10,15, 20-tetraazaporphyrin	−56.0	1.0×10^{-5} 3.5×10^{-1}	2.3–6.4	1.0×10^{-6}	SCN⁻, ClO₄⁻, ClO₃⁻, NO₃⁻	[17]
Cyanoaquacobyric acid heptakis (2-phenylethyl ester)	−49.0	5.0×10^{-5} 1.0×10^{-2}	6.0	-	NO₂⁻	[18]
Substituted onium base salts	−57.6	1.0×10^{-4} 1.0×10^{-1}	7.5–12.0	7.0×10^{-5}	SO_4^{2-}, HCO_3^-, Cl⁻, $H_2PO_4^-$	[19]
Fe^II- and Ni^II-bathophenan-throlineazide ion-pair complexes	−29.2	8.9×10^{-6} 1.0×10^{-1}	4.3–10.5	8.0×10^{-7}	SCN⁻, S^{2-}, NO_2^-, Cl⁻	[20]
Orion ammonium-sensitive gas probe model (95/12) with a Teflon semi-permeable membrane/Teflon membrane	−59.1	1.0×10^{-4} 1.0×10^{-1}	1.0–3.5	3.5×10^{-5}	SO_3^{2-}, NO_2^-, S^{2-}, HCO_3^-, CH_3COO^-	[21]
Orion ammonium-sensitive electrode (model 95/12) with a polypropylene membrane	−58.8	1.0×10^{-4} 1.0×10^{-1}	1.0–3.5	1.9×10^{-5}	SO_3^{2-}, NO_2^-, S^{2-}	[22]
Fe^III-hydrotris-(3,5-dimethyl-pyrazolyl) borate acetylacetonate chloride	−59.4	6.3×10^{-7} 1.0×10^{-2}	3.5–9.0	5.0×10^{-7}	-	[23]
Fe^III- Schiff base	−58.9	1.0×10^{-6} 5.0×10^{-2}	4.3–10.2	8.8×10^{-7}	ClO_4^-, IO_3^-, ClO_4^-, NO_2^-, NO_3^-, Cl⁻, I⁻	[24]
Mn(III)-porphyrin	−56.3	2.2×10^{-5}–1.0×10^{-2}	3.9–6.5	1.3×10^{-5}	I⁻, CN⁻	[25]
Co(II)-phthalocyanine	−48.5	5.1×10^{-5}–1.0×10^{-2}	4.2–6.5	1.7×10^{-5}	SO_3^{2-}	[26]
Mn^II-[2-formyl]quinoline thiosemicarbazone] complex	−55.8	1.0×10^{-5}–1.0×10^{-2}	5.5–9.0	8.0×10^{-6}	-	[26]
Fe-PC/POT	−58.3	3.5×10^{-7}–1.0×10^{-2}	6.0–9.0	1.0×10^{-7}	-	This work
Nit-N₃/POT	−55.1	3.5×10^{-8}–1.0×10^{-2}	5.0–10.0	7.7×10^{-8}	-	This work

2.3. Sensors' Selectivity

Potentiometric selectivity coefficients ($Kpot\ N_3^-, J$) of azide sensors based on nitron-azide (Nit-N_3) and iron (II) phthalocyanine (FePC) plasticized with o-nitrophenyloctyl ether (o-NPOE) were evaluated using the modified separate solutions method (MSSM) [30]. The measured potentiometric selectivity coefficients of Nit-N_3/POT and FePC/POT are given in Table 3. It can be seen that the selectivity coefficients of Nit-N_3/POT based sensor towards different anions were in the order: Sal$^-$ > I$^-$ > ClO$_4^-$ > SCN$^-$ > NO$_3^-$ > NO$_2^-$ > CH$_3$COO$^-$ > Cl$^-$ > Br$^-$ > SO$_4^{2-}$ > PO$_4^{3-}$. This order agreed fairly well with the classical Hofmeister series [1]. However, FePC/POT/based sensor, exhibited enhanced selectivity for azide ion with an anti-Hofmeister order. The selectivity coefficients were in the order: ClO$_4^-$ > I$^-$ > Sal$^-$ > NO$_3^-$ > SCN$^-$ > CH$_3$COO$^-$ > Cl$^-$ > Br$^-$ > NO$_2^-$ > SO$_4^{2-}$ > PO$_4^{3-}$. This selectivity sequence is similar to that reported with some metallophthalocyanine based sensors of other species [24,31,32].

Table 3. Potentiometric selectivity coefficients, *Log $K^{pot}_{x,y}$, of the proposed screen-printed sensors.

Interfering ion	* Log $K^{pot}_{x,y}$	
	[FePC/POT	[Nit/N_3/POT
PO$_4^{3-}$	−6.7 ± 0.2	−6.2 ± 0.5
Salicylate	−3.7 ± 0.4	−0.5 ± 0.1
NO$_2^-$	−5.27 ± 0.5	−3.1 ± 0.3
ClO$_4^-$	−3.3 ± 0.4	−0.8 ± 0.1
SCN$^-$	−4.3 ± 0.7	−1.1 ± 0.1
I$^-$	−3.5 ± 0.4	−0.6 ± 0.2
Cl$^-$	−5.1 ± 0.3	−4.7 ± 0.3
Br$^-$	−5.2 ± 0.1	−4.9 ± 0.3
SO$_4^{2-}$	−5.6 ± 0.4	−5.1 ± 0.4
CH$_3$COO$^-$	−4.9 ± 0.1	−4.2 ± 0.1
NO$_3^-$	−4.2 ± 0.3	−2.3 ± 0.5

* Mean of three measurements.

2.4. Electrochemical Impedance Spectroscopy (EIS) Measurements

EIS measurements were performed on the proposed sensors immersed in 10^{-2} M NaN$_3$ solution. Examples of the impedance spectra of sensors with and without the solid contact (POT) were examined and the results are illustrated in Figure 3 for Nit-N_3/POT, FePC/POT, Nit-N_3 and FePC membranes. The high frequency semicircle can be related to the bulk impedance of the membranes. The bulk resistances (R_b) of Nit-N_3 and FePC membranes were found to be 0.37 and 0.24 MΩ, respectively. This slight difference in R_b values may be attributed to the slight dissolution of POT in the membrane matrix. The time constant (t_b) of the bulk process was calculated from the impedance spectra presented in Figure 3 using the frequency (f_{max}) at the top of the high-frequency semicircle:

$$t_b = (2\pi f_{max})^{-1} \tag{1}$$

For Nit-N_3 and FePC membrane-based sensors, the time constants were 0.32 and 0.25 ms, respectively. In the presence of POT, Nit-N_3/POT and FePC/POT-based membrane sensors displayed smaller time constants of 0.28 and 0.20 ms, respectively. These results suggest that the bulk process of the FePC/POT membrane is faster than that of Nit-N_3/POT membrane. In addition, the bulk process of the sensors became faster with the incorporation of POT.

As shown in Figure 3, the low frequency semicircle in both Nit-N_3/POT and FePC/POT-based sensors is significantly smaller than for the same sensors without a POT layer. This indicates that the charge transfer impedance is decreased by incorporation of POT between the membrane and the solid electrical support. The double-layer capacitance (C_{dl}) of Nit-N_3/POT and FePC/POT membranes were C_{dl} = 22.4 ± 0.7 and 13.8 ± 0.6 µF, respectively, compared with C_{dl} = 4.7 ± 0.2 and 4.2 ± 0.4 µF for Nit-N_3 and FePC membranes, respectively. This further confirm that the presence of POT as an ion-to-electron transducer significantly facilitate faster charge transport between the interfaces and offer more stable potential responses of the solid-contact ion selective sensor.

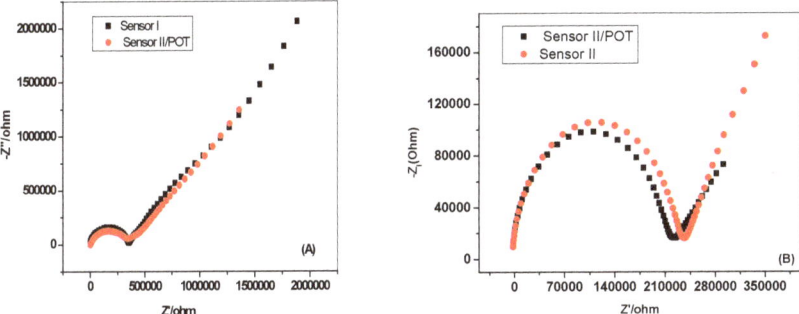

Figure 3. Electrochemical impedance spectroscopy (EIS) measurements of: (**A**) Nit-N$_3$ and (**B**) FePC membrane based sensors.

2.5. Chronopotentiometric Measurements

The potential stability of the proposed sensors was evaluated by using constant current chronopotentiometric measurements [33]. In the absence of POT, Nit-N$_3$ and FePC-based membrane sensors have large potential drift of 181.2 ± 3.1 and 226.0 ± 5.1 (n = 3) µV/s, respectively. However, much less potential drift of about 40.6 ± 1.9 and 69.2 ± 1.1 µV/s (n = 3) was noticed with Nit-N$_3$/POT and Fe-PC/POT based membrane sensors, respectively. This declined potential is attributed to the high double layer capacitance of POT. The capacitances of the sensors were calculated and found to be 24.6 ± 0.8, 14.5 ± 0.7, 5.5 ± 0.3 and 4.42 ± 0.6 µF for Nit-N$_3$/POT, FePC/POT, Nit-N$_3$ and FePC-based membrane sensors, respectively. The data depicted in Figure 4 confirm a clear relationship between the potential stability ($\Delta E/\Delta t$) or the capacitance (C) and the effect of POT solid contact. In addition, there is a good agreement between the results obtained by EIS and chronopotentiometry upon using POT confirming a high compatibility and adhesion between the solid conducting base and the polymeric membrane. This leads to extending the sensor response range, increasing the potential stability and improving the selectivity behavior.

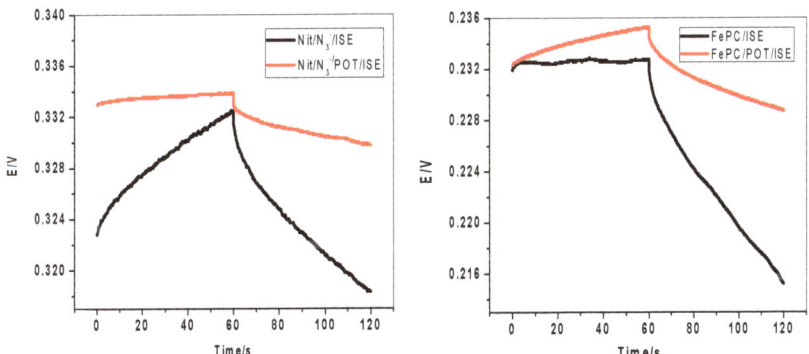

Figure 4. Measurements for azide membrane sensors with and without POT as a solid contact material.

2.6. Effect of Water Film Test of the Electrode Potential

It is well documented that water film formation between polymeric sensing membranes and the solid conducting substrates causes significant potential drift and affect the sensitivity due to poor adhesion and delamination of the polymeric membrane. In the present work, a solid contact layer of poly(octylthiophene) (POT) was used between the solid conductor and polymeric sensing PVC membrane. The sensors were first conditioned in 10^{-2} M N_3^- solution and then the sample was replaced with a solution of 10^{-2} M Na_2SO_4. A control experiment was performed by using Nit-N_3 and FePC PVC membrane-based sensors without POT which are very close to the coated wire CWEs) configuration. As shown in Figure 5, positive EMF changes of ~145 and 200 mV are noted upon replacing 1.0×10^{-5} and 1.0×10^{-4} M N_3^- solution with the electrolytic background solution of Nit-N_3/POT and FePC/POT, respectively.

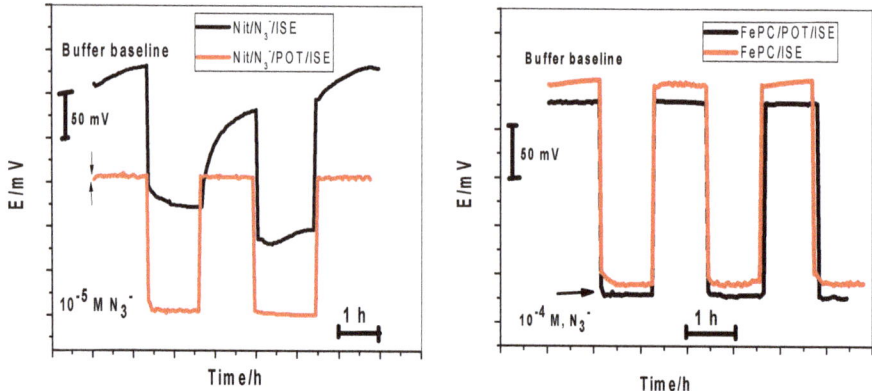

Figure 5. Effect of water-layer on azide membrane sensors with and without POT solid contact.

For Nit-N_3 and FePC, a positive potential drift was noticed. This can be attributed to the formation of water layer between the sensing membranes and the conducting substrate. The stable response of Nit-N_3/POT and FePC/POT-based sensors confirms the absence of a water layer, so the undesirable water layer can be successfully removed by the insertion of the high hydrophobic POT layer between the sensing membrane and the solid conducting substrate.

The long-term response of the proposed sensors was also tested. When not in use, these fabricated sensors were all conditioned in 1.0×10^{-5} M N_3^- solution. Negligible change in the calibration slopes and detection limits was observed during 3 months. These results indicated the absence of the water films. The robust and reliable solid-contact N_3^--ISEs are promising for applications in many fields of contemporary research.

2.7. Analytical Applications

Synthetic primer mixtures containing azide were prepared with azide content matched the real formulations and assessed by the proposed sensors. About 500 mg of $KClO_3$ and 500 mg Sb_2S_3 (stibnite) were mixed with three different accurately measured amounts of sodium azide 1.0, 5.0 and 10.0 mg, respectively. Each mixture was dissolved in a 100 mL measuring flask and completed to the mark with deionized water. The azide contents of these mixtures were determined using the proposed sensors. As shown in Table 4, the analysis of mixtures containing 1.0, 5.0 and 10.0 mg azide showed average recoveries of 98–101%, 102.4–97.4% and 99.3–97.7% ($n = 6$) with the above tested concentrations, respectively.This confirmed the validity of the suggested method for the assessment of azide in complicated matrices.

Table 4. Determination of azide in synthetic primer mixtures using the proposed azide sensors.

Sample	Taken, azide, mg/g	Azide, mg/g[a]			
		[Fe-PC/POT]	Recovery, %	[Nit-N$_3$/POT]	Recovery, %
Mixture 1	1.0	0.98 ± 0.05	98.0	1.01 ± 0.2	101.0
Mixture 2	5.0	5.2 ± 0.4	102.4	4.87 ± 0.1	97.4
Mixture 3	10.0	9.93 ± 0.7	99.3	9.77 ± 0.3	97.7

[a] Average of five measurements ± standard deviation.

3. Materials and Methods

3.1. Apparatus

Screen-printed azide PVC membrane sensors in conjunction with Ag/AgCl double junction reference electrode (model 90-02, Orion, Cambridge, MA, USA, USA) filled with 10% (w/v) K$_2$SO$_4$ in its outer compartment were used for measurement of azide. An Orion pH/meter (model SA 720, Cambridge, MA, USA) and a combination glass pH electrode (Schott blue line 25, Stuttgart, Germany) were used. The cell used for EMF measurements at ambient temperature was of the type: Ag/AgCl/KCl (10^{-2} M)/sample test solution//sensor membrane/POT/C.

Impedance and chronopotentiometric measurements were performed by applying a constant currents of ± 1 nA for 60s, on the screen-printed sensors in presence and absence of poly-(3-octylthiophene) (POT) by using an (Autolab Model 2000) potentiostat/galvanostat (Metrohom Instruments, Herisau, Switzerland). A reference electrode (Ag/AgCl (3 M KCl), and a platinum auxiliary electrode were used. The tested solution was 0.01 M N$_3^-$ ion. Chronopotentiometry was carried out on the proposed sensors by applying a constant currents of ± 1 nA for 60s, respectively. The impedance spectra were measured over the frequency range of 10 kHz to 0.1 Hz. The amplitude of the sinusoidal excitation signal was 50 mV. All experiments were performed at room temperature (23 ± 2 °C).

3.2. Reagents and Materials

All chemicals used were of analytical reagent grade unless stated otherwise, and doubly distilled water was used throughout. o-Nitrophenyloctyl ether (o-NPOE), bis(2-ethylhexyl)phthalate (DOP), dibutylsebacate (DBS), tris-hydroxymethylaminomethane (TRIS), tetradodecylammonium tetrakis(4-chlorophenyl) borate (ETH 500), poly(3-octylthiophene) (POT), tetrahydrofuran (THF), nitron (1,4-diphenylendoanilino-dihydrotriazole; 1,4-diphenyl-3-(phenylamino)-1H-1,2,4-triazole) and iron(II)phthalocyanine (FePC) were purchased from Sigma-Aldrich Chem. Co. (Steinheim, Germany). High relative molecular weight PVC was purchased from Fluka (Buchs, Switzerland). Sodium and potassium salts of all anions were purchased from Merck (Darmstadt, Germany). A stock solution of 0.1 M sodium azide was prepared in pre-boiled doubly distilled water. Working azide standards of different concentration in the range of 10^{-2} to 10^{-9} M were freshly prepared by stepwise dilutions. Tris buffer (0.1 M) of pH 7.0 was used to adjust the pH of all sample solutions. The ion activity coefficients were calculated according to the Debye–Hückel equation.

3.3. Nitron-azide Ion-pair Complex

A 10^{-2} M of nitron solution was prepared by dissolving 0.31 g of nitron in 100 mL of 20% acetic acid. A 20 mL portion of 10^{-2} M nitron solution was mixed with a 10 mL aliquot of 10^{-2} M NaN$_3$. The solution was thoroughly mixed and stirred for 15 min. A brown precipitate was formed, filtered off, washed with bidistilled water, dried at room temperature and ground to a fine powder in agate mortar. Elemental analyses of the precipitate confirm the formation of 1:1 (azide: nitron) ion association complex.

3.4. Sensor Fabrication

Screen-printed ceramic carbon sensor substrate (96% alumina) was purchased from Gwent Electronic Materials Ltd. (Lancster, UK). A carbon electrode with an area of 3.1 mm and a diameter of 2 mm was used as a working electrode. The azide sensors were prepared by NaN_3 through two steps; the first involved fabrication of the solid contact conducting substrate and the second dealt with screen printing of electroactive materials [34–36]. Poly(3-octylthiophene) (POT) in chloroform solution (2.0 mM) was successively drop-cast four times on the carbon-based contact and the chloroform solvent was allowed to evaporate after formation of each layer. The recognition membrane cocktail was prepared by mixing a 2.5 mg portion (2.5 wt %) of the ion sensing material (Nit-N_3 or FePC), 2.0 mg (2.0% wt %) of ETH-500, 32.2 mg (32.2 wt %) of poly(vinylchloride) (PVC) and 63.3 mg (63.3 wt %) of *o*-NPOE and dissolving in 2 mL of THF. A 10 µL aliquot of the membrane cocktail was added dropwise over the POT layer through the circular orifice of the screen-printed wafer (SP) and left 4 h for drying at room temperature in the dark. A schematic representation of the solid-contact ion selective micro sensor is shown in Figure 1. The sensors were conditioned before use by soaking in a 1.0×10^{-4} M aqueous N_3^- solution, and were stored in the same solution when not in use.

3.5. Sensor Calibration and Selectivity Measurements

The (FePC/POT) and (Nit-N_3/POT)-based membrane sensors were calibrated by immersion, along with an Ag/AgCl reference electrode, in a 50-mL beaker containing 9.0 mL of 10^{-2} M Tris buffer solution of pH 7.0. Aliquots (1.0 mL) of standard sodium azide solution (1.0×10^{-7} to 1.0×10^{-1} M) were successively added with continuous stirring. The potential readout was recorded for each solution after stabilization to ± 0.5 mV (2 min). Calibration graphs connecting potential reading with logarithm azide concentrations were plotted and used for all subsequent unknown azide measurements. Selectivity coefficients were determined using the modified separate solutions method by recording separate calibration curves for all the interfering ions of interest [30]. The selectivity values were determined from the highest measured concentrations (0.1 M) with established formalisms.

4. Conclusions

Solid contact carbon screen-printed ceramic azide micro-sensors were developed, electrochemically characterized and used. These sensors are based on the utilization of ironII-phthalocyanine (Fe-PC) neutral carrier and nitron-azide ion-pair complex (Nit-N_3^-) as selective recognition receptors, POT as a solid contact material on a carbon printed ceramic substrate. The sensors displayed extended linear response range (1.0×10^{-2}–1.0×10^{-7} M), low detection limit (1.0×10^{-7} and 7.7×10^{-8} M) and fast response time (< 10 s) for FePC/POT and Nit-N_3/POT, membrane-based sensors, respectively. The potential sensitivity and stability of these sensors were tested by electrochemical impedance spectroscopy (EIS) and constant-current chronopotentiometry techniques. The proposed solid-contact azide-sensors were successfully used for trace azide quantification. The sensors offered good advantages over many of those previously described in terms of robustness, ease of fabrication, potential stability, selectivity, and accuracy. The sensors can be introduced in a flow system for contineous azide monitoring. No sample pretreatment is requried for azide analysis using these proposed sensors.

Author Contributions: The listed authors contributed to this work as described in the following: A.G.E., A.H.K., and S.S.M.H. gave the concepts of the work, interpreted the results, the experimental part and prepared the manuscript, A.H.K. and S.S.M.H. cooperated in the preparation of the manuscript and A.H.K., A.E.-G.E.A. and S.S.M.H. performed the revision before submission. A.E.-G.E.A. revealed the financial support for the work. All authors read and approved the final manuscript.

Funding: The authors are grateful to the Deanship of Scientific Research, king Saud University for funding through Vice Deanship of Scientific Research Chairs.

Conflicts of Interest: The authors declare no conflict of interest.

References

1. Cuartero, M.; Crespo, G.A. All-solid-state potentiometric sensors: A new wave for in situ aquatic research. *Curr. Opin. Electrochem.* **2018**, *10*, 98–106. [CrossRef]
2. Mikhelson, K.N. *Ion-Selective Electrodes*; Springer: Berlin/Heidelberg, Germany, 2013.
3. Bieg, C.; Fuchsberger, K.; Stelzle, M. Introduction to polymer-based solid-contact ion-selective electrodes—Basic concepts, practical considerations, and current research topics. *Anal. Bioanal. Chem.* **2017**, *409*, 45–61. [CrossRef] [PubMed]
4. Cadogan, A.; Gao, Z.; Lewenstam, A.; Ivaska, A.; Diamond, D. All-solid state sodium-selective electrode based on a calixareneionophore in a poly(vinyl chloride) membrane with a polypyrrole solid contact. *Anal. Chem.* **1992**, *64*, 2496–2501. [CrossRef]
5. Hu, J.; Stein, A.; Buhlmann, P. Rational design of all-solid-state ion-selective electrodes and reference electrodes. *Trends Anal. Chem.* **2016**, *76*, 102–114. [CrossRef]
6. Sokalski, T.; Ceresa, A.; Zwickl, T.; Pretsch, E. Large improvement of the lower detection limit of ion-selective polymer membrane electrodes. *J. Am. Chem. Soc.* **1997**, *119*, 11347–11348. [CrossRef]
7. Bakker, E.; Meyerhoff, M.E. Ionophore-based membrane electrodes: New analytical concepts and non-classical response mechanisms. *Anal. Chim. Acta* **2000**, *416*, 121–137. [CrossRef]
8. Bakker, E.; Pretsch, E. Potentiometry at trace levels. *Trends Anal. Chem.* **2001**, *20*, 11–19. [CrossRef]
9. Bakker, E.; Bhakthavatsalam, V.; Gemene, K.L. Beyond potentiometry: Robust electrochemical ion sensor concepts in view of remote chemical sensing. *Talanta* **2008**, *75*, 629–635. [CrossRef] [PubMed]
10. Renedo, O.D.; Alonso-Lomillo, M.A.; Martinez, M.J.A. Recent developments in the field of screen-printed electrodes and their related applications. *Talanta* **2007**, *73*, 202–219. [CrossRef] [PubMed]
11. Li, P.; Liang, R.; Yang, X.; Qin, W. Imprinted nanobead-based disposable screen-printed potentiometric sensor for highly sensitive detection of 2-naphthoic acid. *Mater. Lett.* **2018**, *225*, 138–141. [CrossRef]
12. Carroll, S.; Baldwin, R.P. Self-calibrating microfabricated iridium oxide pH electrode array for remote monitoring. *Anal. Chem.* **2010**, *82*, 878–885. [CrossRef]
13. Betterton, E.A. Environmental fate of sodium azide derived from automobile air bags. *Crit. Rev. Environ. Sci. Technol.* **2003**, *33*, 423–458. [CrossRef]
14. Howard, J.D.; Skorgeboe, K.J.; Case, G.A.; Raysis, V.A.; Laqsina, E.Q. Death following accidental sodium azide investigation. *J. Forensic Sci.* **1990**, *35*, 193–196. [CrossRef]
15. Kleinhofs, A.; Owais, W.M.; Nilan, R.A. Azide. *Mutat. Res.* **1978**, *55*, 165–195. [CrossRef]
16. Chang, S.; Lamm, S.H. Human health effects of sodium azide exposure a literature review and analysis. *Int. J. Toxicol.* **2003**, *22*, 175–186. [CrossRef]
17. Prasad, R.; Gupta, V.K.; Kumar, A. Metallo-tetraazaporphyrin based anion sensors: Regulation of sensor characteristics through central metal ions coordination. *Anal. Chim. Acta* **2004**, *508*, 61–70. [CrossRef]
18. Watanabe, K.; Noguchi, O.; Okada, K.; Katsu, T. An azide-sensitive electrode for criminal investigation. *Jpn. J. Forensic Toxicol.* **1999**, *17*, 180–186.
19. Saville, B. A Scheme for the colorimetric determination of microgram amounts of thiols. *Analyst* **1958**, *83*, 670–672. [CrossRef]
20. Hassan, S.S.M.; ElZawawy, F.M.; Marzouk, S.A.M.; Elnemma, E.M. Polyvinyl chloride matrix membrane electrodes for manual and flow injection determination of metal azides. *Analyst* **1992**, *17*, 1683–1689. [CrossRef]
21. Van, J.T.; vanden Berg, C.M.G.; Martin, T.D. Gas electrode method for determination of azide in aqueous samples from there processing industry. *Anal. Commun.* **1997**, *34*, 385–387.
22. Hassan, S.S.M.; Ahmed, M.A.; Marzouk, S.A.M.; Elnemma, E.M. Potentionetric gas sensor for the selective determination of azides. *Anal. Chem.* **1991**, *63*, 1547–1552. [CrossRef]
23. Singh, A.K.; Singh, U.P.; Aggarwal, V.; Mehtab, S. Azide-selective sensor based on tripodal iron complex for direct azide determination in aqueous samples. *Anal. Bioanal. Chem.* **2008**, *391*, 2299–2380. [CrossRef]
24. Ghaedi, M.; Shokrollahi, A.; Montazerozohori, M.; Derki, S. Design and construction of azide carbon paste selective electrode based on a new schiff's base complex of iron. *IEEE Sens. J.* **2010**, *10*, 814–819. [CrossRef]
25. Hassan, S.S.M.; Kelany, A.E.; Al-Mehrezi, S.S. Novel polymeric membrane sensors based on Mn(III) porphyrin and Co(II) phthalocyanineionophores for batch and flow injection determination of azide. *Electroanalysis* **2008**, *20*, 438–443. [CrossRef]

26. Kamel, A.H. New potentiometric transducer based on a Mn(II)[2-formylquinoline thiosemicarbazone] complex for static and hydrodynamic assessment of azides. *Talanta* **2015**, *144*, 1085–1090. [CrossRef]
27. Lindner, E.; Gyurcsanyi, R.E. Quality control criteria for solid-contact, solvent polymeric membrane ion-selective electrodes. *J. Solid State Electrochem.* **2009**, *13*, 51–68. [CrossRef]
28. Moreira, F.T.C.; Guerreiro, J.R.L.; Azevedo, V.L.; Kamel, A.H.; Sales, M.G.F. New biomimetic sensors for the determination of tetracycline in biological samples: Batch and flow mode operations. *Anal. Methods* **2010**, *2*, 2039–2045. [CrossRef]
29. Perez, M.A.A.; Marın, L.P.; Quintana, J.C.; Pedram, M.Y. Influence of different plasticizers on the response of chemical sensors based on polymeric membranes for nitrate ion determination. *Sens. Actuators B* **2003**, *89*, 262–268. [CrossRef]
30. Bakker, E. Determination of improved selectivity coefficients of polymer membrane ion-selective electrodes by conditioning with a discriminated ion. *J. Electrochem. Soc.* **1996**, *143*, L83–L85. [CrossRef]
31. Hassan, S.S.M.; Marzouk, S.A.M.; Mohamed, A.H.K.; Badawy, N.M. Novel dicyanoargentate polymeric membrane sensors for selective determination of cyanide ions. *Electroanalysis* **2004**, *16*, 298–303. [CrossRef]
32. Almeer, S.H.M.A.; Zogby, I.A.; Hassan, S.S.M. Novel miniaturized sensors for potentiometric batch and flow-injection analysis (FIA) of perchlorate in fireworks and propellants. *Talanta* **2014**, *129*, 191–197. [CrossRef]
33. Bobacka, J. Potential Stability of All-Solid-State Ion-Selective Electrodes Using Conducting Polymers as Ion-to-Electron Transducers. *Anal. Chem.* **1999**, *71*, 4932–4937. [CrossRef]
34. Kamel, A.H.; Galal, H.R.; Awwad, N.S. Cost-effective and handmade paper-based potentiometric sensing platform for piperidine determination. *Anal. Methods* **2018**, *10*, 5406–5415. [CrossRef]
35. Abdalla, N.S.; Youssef, M.A.; Algarni, H.; Awwad, N.S.; Kamel, A.H. All Solid-State Poly (Vinyl Chloride) Membrane Potentiometric Sensor Integrated with Nano-Beads Imprinted Polymers for Sensitive and RapidDetection of Bispyribac Herbicide as Organic Pollutant. *Molecules* **2019**, *24*, 712. [CrossRef]
36. Amr, A.E.; Al-Omar, M.A.; Kamel, A.H.; Elsayed, E.A. Single-Piece Solid Contact Cu^{2+}-Selective Electrodes Based on a Synthesized Macrocyclic Calix [4]arene Derivative as a Neutral Carrier Ionophore. *Molecules* **2019**, *24*, 920. [CrossRef]

Sample Availability: Not Available.

© 2019 by the authors. Licensee MDPI, Basel, Switzerland. This article is an open access article distributed under the terms and conditions of the Creative Commons Attribution (CC BY) license (http://creativecommons.org/licenses/by/4.0/).

MDPI
St. Alban-Anlage 66
4052 Basel
Switzerland
Tel. +41 61 683 77 34
Fax +41 61 302 89 18
www.mdpi.com

Molecules Editorial Office
E-mail: molecules@mdpi.com
www.mdpi.com/journal/molecules

www.ingramcontent.com/pod-product-compliance
Lightning Source LLC
LaVergne TN
LVHW070625100526
838202LV00012B/729